LAW
FOR
PROFESSIONAL
ENGINEERS

Third Edition

Canadian and International Perspectives

LAW FOR PROFESSIONAL ENGINEERS

Third Edition

Canadian and International Perspectives

D.L. Marston, B.Sc., P.Eng., LL.B.
of Osler, Hoskin & Harcourt
Barristers & Solicitors
Toronto
and
Special Lecturer in Engineering Law
Faculty of Applied Science and Engineering
University of Toronto

McGraw-Hill Ryerson Limited

Toronto   New York   Auckland   Bogotá   Caracas
Lisbon   London   Madrid   Mexico   Milan   New Delhi
San Juan   Singapore   Sydney   Tokyo

LAW FOR PROFESSIONAL ENGINEERS
Canadian and International Perspectives
Third Edition

CAN©OPY

ISBN: 0-07-552628-X

3 4 5 6 7 8 9 10   RRD   5 4 3 2 1 0 9 8 7

Printed and bound in the United States of America

Care has been taken to trace ownership of copyright material contained
in this text. The publisher will gladly take any information that will
enable them to rectify any reference or credit in subsequent editions.

Editor-in-Chief: Dave Ward
Developmental Editor: Laurie Graham
Production Editor: Gail Marsden
Production Co-ordinator: Nicla Dattolico
Cover Design: Samantha Taylor
Typesetter: McGraphics Desktop Publishing Limited
Printer: R.R. Donnelly & Sons Company

Canadian Cataloguing in Publication Data

Marston, D.L. (Donald L.), date
    Law for professional engineers

3rd ed.
Includes index.
ISBN 0-07-552628-X

1. Engineering law – Canada.   2. Engineering contracts – Canada.
3. Engineering – Legal status, laws, etc. – Canada.   I. Title.

KE2730.M37   1996   346.71'0024624   C95–933275-8
KF2928.M37   1996

TABLE OF CONTENTS

PREFACE

Consistent with the purpose of the first edition in 1981, the objective of this updated and expanded third edition is to provide a current law text to serve the Canadian engineering profession.

In the ten years that have passed since the second edition, the practice of professional engineering continues to be affected by new legislation and relevant court decisions. In addition, the interest of governments in privatizing energy, water, and transportation projects, in Canada and abroad, has led to new opportunities and challenges for engineers. These developments have generated keen interest in public/private partnerships and build-operate-transfer ("BOT") projects, in which engineers play a major role. The expanding scope of risks and responsibilities undertaken on infrastructure projects increases the need for engineers to understand the importance of appropriate attention to legal issues.

This third edition includes new commentary on public/private partnerships, BOT projects, alternative dispute resolution ("ADR"), and risk assessment and related issues. It also includes case law developments of the past ten years in areas such as bid protests, concurrent liability in tort and contract, tort remedies in construction, limitation periods, fundamental breach, and the liability of engineers as employees. Much has happened in Canada that relates to engineers.

Much has also happened that has drawn more and more Canadian engineers, contractors, and manufacturers into the international marketplace. In response to expressions of interest from Canadian engineers, this edition also addresses selected international legal issues. Commentary has been included on organizational considerations in foreign jurisdictions, international performance bonds and guarantees, risk issues on international infrastructure projects, and the North American Free Trade Agreement. Hopefully, this edition will assist engineers with their international plans and initiatives.

This book focuses on fundamentals and legal principles of relevance to engineers. Certain statutory examples place emphasis on Ontario legislation, but similar statutes of other common-law provinces are generally referenced. A chapter on the Law of Quebec as relevant to engineers has also been included. It is important that Canadian engineers are aware of the unique nature within Canada of the civil law system in the Province of Quebec.

Although broad in scope, this is not an exhaustive law text. It deals with selected areas of the law. It is not a substitute for appropriate legal advice on specific matters. Such advice is highly recommended in the interests of avoiding problems through preventive planning and contracting and is essential should legal problems arise.

I am grateful for suggestions and recommendations received prior to preparing this third edition from fellow lecturers, including David Pratt of the Southern Alberta Institute of Technology, Richard Rennie of the University of Victoria, and Bryan Shapiro of the University of British Columbia.

I am also grateful for continuing insights into relevant issues provided by Professional Engineers Ontario and, in that regard, I want to particularly thank Harry H. Angus, P. Eng., Peter Large, P. Eng., and Debra Dileo, P. Eng. I also wish to thank those of my many Canadian and international partners and colleagues who have assisted in providing insights, and particularly those who assisted in updating aspects of this text. I owe a special thanks to my friend and colleague Carlo Greco, P. Eng., LL.B. for his assistance in adding to the scope of the appendix of sample case studies.

I am most grateful to my colleague Olivier F. Kott, LL. L. of Ogilvy Renault in Montreal, a Quebec lawyer who has contributed important insights into legal issues of relevance to engineers in the Province of Quebec by writing the chapter in this book on the Law of Quebec.

I particularly want to thank my partners D. Robert Beaumont, P. Eng., LL.B. and Michael Gough, LL.M. with whom I have the pleasure of closely sharing a Canadian and international law practice that generates many of the project insights reflected in this book.

Donald L. Marston
August, 1995

CHAPTER ONE

THE CANADIAN LEGAL SYSTEM

HISTORICAL BASIS

The legal system of the nine common-law provinces and the territories of Canada is based upon the English common-law system. It is important to understand something of the evolution of the English system in order to appreciate how the Canadian system operates.

At one of its early stages of development, the English legal system was very rigid. Certain specific remedies were available in only certain circumstances. This system of specific remedies was called the "common law." As time passed, it became evident that the specific remedies provided by the common-law courts were not sufficient. Where relief beyond the scope of the common law was sought, special appeals were made to the English monarch; if the monarch saw fit to exercise his or her discretion, a remedy more "equitable" than that provided by the common law was declared. Eventually, the English "courts of equity" were developed as a separate court system, providing more reasonable remedies as circumstances required. As pointed out in *Black's Law Dictionary*, the term "equity," in its broadest and most general sense, denotes a spirit of fairness, justness, and right dealing ... grounded in the precepts of conscience.

Eventually, the two systems — the old common-law system and the courts of equity — were combined, and an improved system was developed to provide remedies premised on both common-law precedents and on equitable principles. This improved system continued to be called the "common law," and is the system from which the present Canadian common-law system evolved.

THE THEORY OF PRECEDENT

In deciding cases, the courts apply legal principles established in previous court decisions that involved similar or analogous fact situations; this is called "the theory of precedent." But the courts also dispense equitable relief and thus there is flexibility in the courts' decision-making process. At times, to slavishly follow precedent would not reflect society's values: hence a court can exercise its equitable discretion to reach a policy decision that may represent a departure from case precedents.

Factual distinctions between cases may also provide the basis for flexibility. A court may see fit to dismiss the application of a precedent on the basis of relatively minor factual distinctions between the precedent and the facts of the case before the court, provided the end result is justified.

However, departures from established precedents are often very slow to evolve. This slow evolution is a characteristic of our legal system that may, at times, be criticized; nevertheless, the theory of precedent is of major importance and is the basis of predictability in the legal system.

THE COMMON LAW

A major source of law is "judge-made law" — court decisions establishing legal principles.

LEGISLATION

In addition to the "common law" or "judge-made law," an extremely important source of law is legislation: statutes enacted by elected legislatures. A statute is a codification of the law as the legislature determines at the time of enactment; it may be codification of existing common law or the enactment of new law.

Applicability of a statute can be questioned in a lawsuit; if so, it is up to the court to determine whether the statute does apply to the facts of the case. The court must apply the statute appropriately. At times, the wording of a statute must be interpreted by the court.

Some statutes provide for regulations as a further source of law. The Professional Engineers Act of Ontario[1] (and similar statutes of other common-law provinces, for example) provides

---

[1]   R.S.O. 1990 c. P-28
(Note that, in the interests of brevity, statute citations throughout the text do not include amendments.)

that an elected and appointed council may prescribe the scope and conduct of examinations of candidates for registration; and may regulate other matters, such as the designation of specialists.

When made in accordance with an authorizing statute, regulations are another source of law.

Many statutes are relevant to the professional engineer. It is important that the engineer complies with federal and provincial statutes of relevance to his or her practice and that the engineer is aware of amendments and new statutes.

FEDERAL AND PROVINCIAL POWERS

Under the Canadian Constitution, the British North America Act, 1867 (renamed the Constitution Act, 1867 by the Constitution Act, 1982), the federal government and the provinces have authority to enact legislation. The division of powers between the federal government and the provinces is expressed in Sections 91 and 92 of the Constitution Act, 1867, excerpts from which are reproduced, for illustrative purposes, as follows:

91. It shall be lawful for the Queen, by and with the Advice and Consent of the Senate and House of Commons, to make Laws for the Peace, Order, and good Government of Canada, in relation to all Matters not coming within the Classes of Subjects by this Act assigned exclusively to the Legislatures of the Provinces; and for greater Certainty, but not so as to restrict the Generality of the foregoing Terms of this Section, it is hereby declared that (notwithstanding anything in this Act) the exclusive Legislative Authority of the Parliament of Canada extends to all Matters coming within the Classes of Subjects next herein-after enumerated; that is to say, —

Legislative Authority of Parliament of Canada

2. The Regulation of Trade and Commerce....

3. The raising of Money by any Mode or System of Taxation....

10. Navigation and Shipping....

15. Banking, Incorporation of Banks, and the Issue of Paper Money....

21. Bankruptcy and Insolvency.

22. Patents of Invention and Discovery.

23. Copyrights....

27. The Criminal Law...

29. Such Classes of Subjects as are expressly excepted in the Enumeration of the Classes of Subjects by this Act assigned exclusively to the Legislatures of the Provinces.

And any Matter coming within any of the Classes of Subjects enumerated in this Section shall not be deemed to come within the Class of Matters of a local or private Nature comprised in the Enumeration of the Classes of Subjects by this Act assigned exclusively to the Legislatures of the Provinces.

Exclusive Powers of Provincial Legislatures.

Subjects of Exclusive Provincial Legislation.

92. In each Province the Legislature may exclusively make Laws in relation to Matters coming within the Classes of Subject next herein-after enumerated; that is to say, — ....

2. Direct Taxation within the Province in order to the raising of a Revenue for Provincial Purposes....

5. The Management and Sale of the Public Lands belonging to the Province and of the Timber and Wood thereon....

10. Local Works and Undertakings other than such as are of the following Classes: —
    (a) Lines of Steam or other Ships, Railways, Canals, Telegraphs, and other Works and Undertakings connecting the Province with any other or others of the Provinces, or extending beyond the Limits of the Province;
    (b) Lines of Steam Ships between the Province and any British or Foreign Country;
    (c) Such Works as, although wholly situate within the Province, are before or after their Execution declared by the Parliament of Canada to be for the general Advantage of Canada or for the Advantage of Two or more of the Provinces.

11. The Incorporation of Companies with Provincial Objects....

13. Property and Civil Rights in the Province.

14. The Administration of Justice in the Province, including the Constitution, Maintenance, and Organization of Provincial Courts, both of Civil and of Criminal

Jurisdiction, and including Procedure in Civil Matters in those Courts....

16. Generally all Matters of a merely local or private Nature in the Province.

Note that Section 92 of the Constitution Act, 1867 grants to the provinces certain exclusive powers. Section 91, on the other hand, enumerates specific matters that fall within the exclusive legislative authority of the Parliament of Canada. It also provides that the federal parliament shall have authority "to make laws for the Peace, Order, and good Government of Canada" with respect to matters that are not within the exclusive authority of the provinces. This general reference to "Peace, Order, and good Government" is open to broad interpretation, providing a basis for extensive federal legislative powers where circumstances may raise concerns of national importance.

The provincial legislatures are generally empowered to enact statutes dealing with matters of a provincial nature, including property rights. Mechanics', construction, or builders' lien legislation is an example of provincial statute law that may be of particular importance to the engineer.

At times, a dispute may arise as to who has the authority to enact a statute — the federal government or a provincial legislature. Traditionally, if a party sought to challenge the "constitutionality" of a statute, the party had to convince a court that the statute was beyond the authority of the government that enacted it. Provincial statutes would be challenged on the basis that they dealt with a matter assigned to the federal Parliament in section 91 of the Constitution Act, 1867 (for example, the regulation of trade and commerce, criminal law). Federal statutes would be attacked on the grounds that they dealt with a matter within a province's jurisdiction (for example, local works, property and civil rights in a province). If a court could be convinced that the statute was beyond such authority, or "ultra vires," that statute would be effectively rendered void.

Before the Canadian Charter of Rights and Freedoms was enacted as part of the Constitution Act, 1982, courts could strike down legislation on the basis that it was beyond the authority of the government that passed it. But the courts could not void a law because it offended civil liberties. Canadian courts had adopted the British "doctrine of Parliamentary Supremacy." This meant that, theoretically, the Courts could not question the wisdom of any statute, even one that offended civil liberties. Although a court might find that a particularly offensive piece of legislation was beyond the authority of one level of government, that same legislation would be within the authority of another level of government. Theoretically, then, all the rights of the citizens of Canada could be removed by the government if it so desired. The only question would be which of the governments, federal or provincial, would have the ability to remove each right.

This has been dramatically changed through the enactment of the Canadian Charter of Rights and Freedoms as part of the Constitution Act, 1982. The Charter provides that everyone has certain rights. For example, section 2 provides that:

> Everyone has the following fundamental freedoms:
>
> (a) freedom of conscience and religion;
> (b) freedom of thought, belief, opinion and expression, including freedom of the press and other media of communication;
> (c) freedom of association.

The Charter is particularly significant because section 52(1) provides:

> The Constitution of Canada is the supreme law of Canada, and any law that is inconsistent with the provisions of the Constitution is, to the extent of the inconsistency, of no force or effect.

Courts now have the power to rule that statutes are invalid because they violate rights guaranteed by the Charter. The Charter has reduced the effect of the doctrine of Parliamentary Supremacy by placing some limits on the powers of Parliament and the provincial legislatures.

However, it is important to realize that the Charter does not completely eliminate the concept of Parliamentary Supremacy. Section 1 provides, in effect, that governments may, through statutes, place "reasonable limits" on the rights outlined in the Charter. However, if called upon, the government will have to show that such limits are necessary.

Further, Section 33 of the Charter provides that a government may expressly override certain provisions of the Charter. That is, the government may declare that a statute is valid even if the statute violates certain Charter rights. The theory is that a government will be reluctant to announce to its citizens that it believes it is violating the rights of the people. Accordingly, it is assumed that the Section 33 override power will be used sparingly.

THE FEDERAL AND PROVINCIAL COURT SYSTEMS

The most persuasive precedent is usually the decision of most recent date from the highest court. Decisions of the Supreme Court of Canada rank highest, followed by decisions of the Court of Appeal of the province in which a case is commenced. Precedents from other common-law jurisdictions may also be followed. For example, where a provincial court cannot follow a precedent from a higher court within the province or from the Supreme Court of Canada, it may follow a precedent set by the courts of another province, or by a court in another common-law jurisdiction. England and the United States provide common-law precedents; Canadian courts have more often preferred to follow English case precedents than U.S. common law.

The court system within each of the common-law provinces is generally the same; the system was modelled on the English system, and consists of a number of different province-wide courts responsible for a variety of matters. For example, the Ontario Court of Justice is composed of two divisions, the General Division and the Provincial Division. The General Division deals with large claims and federal criminal matters. Small Claims Court falls under a branch of the General Division and deals with disputes involving relatively small amounts of money. The Provincial Division deals with domestic matters (except divorce) and criminal matters involving provincial offences. The province is divided into eight regions. Each region has a regional senior judge for each of the General and Provincial Divisions who manages the judicial resources in the region. The Ontario Court of Appeal is the final Court of Appeal for the province.

Federal courts include The Federal Court of Canada, which has jurisdiction over federal matters such as patents, trade-marks, and copyright, and the Supreme Court of Canada, Canada's final appeal court.

PUBLIC AND PRIVATE LAW

Certain areas of the law are often classified as either public or private law.

Public law deals with the rights and obligations of government, on the one hand, and individuals and private organizations, on the other. Examples of public law are criminal law and constitutional law. Private law deals with rights and obligations of individuals or private organizations. Examples of private law that will be discussed in this text are contracts and torts.

THE LAW OF QUEBEC

Quebec's legal system is not founded upon the English common-law system. The civil law or Civil Code of Quebec has evolved from the Napoleonic Code, and is different from the common-law system. A summary of the Law of Quebec, written by a very experienced Quebec lawyer, Olivier Kott of the Ogilvy Renault firm in Montreal, is set out in Chapter 34 of this text. It is most important that engineers doing business in the Province of Quebec seek legal advice from Quebec counsel.

Reference to the civil law of Quebec can create confusion, as the term "civil" is also used in our common-law system. In the common-law system, "civil" usually means "private." "Civil litigation," for example, refers to a dispute under *private* law, rather than a *criminal* law dispute.

BASIC TERMINOLOGY

In order to appreciate references in this text, an understanding of some basic terminology is important.

(a) Litigation — a lawsuit

(b) Plaintiff — In civil litigation: the party bringing the action or making the claim in the lawsuit. In criminal matters, the "plaintiff" is usually the Crown.

(c) Defendant — The party defending the action, or the party against whom the claim has been made. In criminal matters, the "defendant" is called the "accused."

(d) Appellant — The party appealing the decision of a lower Court, in either civil litigation or criminal matters.

(e) Respondent — The party seeking to uphold a decision of the lower Court that is being appealed. The term applies in both civil litigation and criminal matters.

(f) Privity of contract — Describes the legal relationship between parties to a contract.

(g) Creditor — A party to whom an amount is owing.

(h) Debtor — A party that owes an amount to a creditor.

(i) Indemnification — A promise to directly compensate or reimburse another party for a loss or cost incurred. An indemnification, or "indemnity," is similar to a guarantee; the essential difference is that indemnity rights can be exercised directly. For example, a guarantee works as follows: suppose Jason Smith promises John Doe that Smith will guarantee the debts of ABC Corporation. Enforcement of the guarantee requires that ABC Corporation defaults in making its payment to John Doe and that John Doe first looks to ABC Corporation for such payment. Only then may Doe require payment from the guarantor, Jason Smith. An indemnification, however, works as follows: suppose Jason Smith had indemnified John Doe on account of ABC Corporation's indebtedness to him. As soon as any such debt is incurred, John Doe can require payment directly from Jason Smith without first pursuing ABC Corporation for payment. (As a practical matter, it is often very difficult to distinguish a guarantee from an indemnity, as the guarantee may, for example, by its terms cover the essential difference as described above. That is, a guarantee may expressly provide that the creditor need not exhaust his or her remedies against a debtor before pursuing the guarantor for payment.)

CHAPTER TWO

BUSINESS ORGANIZATIONS

BASIC FORMS

An awareness of the three basic forms of business organizations — sole proprietorships, partnerships, and corporations — is essential to the engineer's appreciation of legal rights and liabilities.

In a sole proprietorship, as the name suggests, an individual carries on business by and for himself or herself. The proprietor personally enjoys the profits of the enterprise and personally incurs any business losses of the enterprise.

A partnership is an association of persons who conduct a business in common with a view to profit. Individuals or organizations carrying on business in partnership share profits and losses personally. One presumed advantage of partnership is that there is strength in numbers, and the combining of energies and talents of individuals may well be advantageous. The essential risk of partnership, however, is that the partnership may incur substantial debts, which the partnership business is unable to pay, with the result that the partnership's creditors may obtain judgments against the partners personally. Such judgments are sometimes satisfied only by seizure and sale of the partners' personal assets — a grim possibility.

Unlike the sole proprietorship and the partnership, the corporation is an entity unto itself, distinct from its shareholder owners. The corporation as an entity has been described as a "fictitious person." The corporation itself owns its assets and incurs its own liabilities; it can sue or be sued in its own name. In fact, a shareholder of a corporation can contract with or sue that corporation.

THE INDEPENDENCE OF
THE CORPORATE ENTITY

The existence of a corporation as separate and apart from its shareholder-owners, and the basic premise that a corporation's liabilities are its own and not those of its shareholders, has long been recognized by our courts. This separate existence provides a strong incentive for individuals to incorporate rather than carry on as sole proprietors or as partners, as the personal assets of sole proprietors and partners remain vulnerable to business creditors.

The courts' recognition of the separate status of the corporation was confirmed in the 1897 decision of the English House of Lords in *Salomon* v. *Salomon & Co. Ltd.*[1] Salomon had, for many years, carried on business as a leather merchant and wholesale boot manufacturer. Eventually he incorporated a company, to which he sold his business. The shareholders of the company consisted of himself and his family, and Salomon held the majority of the shares personally. As part payment of the purchase price for the sale of his business to the corporation, shares of the corporation were issued to Salomon; in addition, debentures constituting security, to evidence the corporation's indebtedness, were also issued to him. All of the requirements of the governing corporate statute were complied with, and at the date of the sale the business was solvent. Eventually, however, the business experienced difficult times and went into insolvency. A lawsuit resulted. The issue was whether Salomon ranked before the general unsecured creditors of the corporation by virtue of being a secured debenture holder. The English Court of Appeal was of the opinion that the incorporation and the sale of the business was a scheme to enable Salomon to carry on business in the name of the corporation with limited liability. The Court of Appeal also thought that Salomon had been trying to obtain a preference over unsecured creditors of the company. However, on appeal, the House of Lords recognized Salomon's corporation as a separate and distinct entity from himself. The House of Lords emphasized that there was no evidence of intent by Salomon to deceive or defraud.

But where it can be established that the limited-liability characteristic of a corporation is being used for the protection of an individual in perpetrating a fraud, the courts will refuse to recognize the separate identities of the individual and the corporation.

[1]   [1897] A.C. 22

To illustrate: in the 1972 decision in *Fern Brand Waxes Ltd.* v. *Pearl*,[2] the Ontario Court of Appeal determined that the defendant, who was a director, officer, and accountant of the plaintiff, Fern Brand Waxes Ltd., had used his position to transfer unauthorized funds. He used the funds as a loan, which was made by the plaintiff to two companies controlled by the defendant. Part of those transferred funds were used to pay for shares in Fern Brand Waxes Ltd. The court determined that such payment for the shares of Fern Brand Waxes Ltd. did not constitute a proper payment for such shares. The court stated that the defendant should not be allowed to profit from his breach of trust. In such circumstances, the corporate character of his two companies was no shield for his conduct, because each company was his instrument and was used to divert funds for his own purposes.

There are other exceptional circumstances, short of fraud, where the courts will intervene to "lift the corporate veil." One such case was *Nedco Ltd.* v. *Clark et al.*,[3] a 1973 decision of the Saskatchewan Court of Appeal. Nedco Ltd. was a wholly owned subsidiary of Northern Electric Company Limited. Employees of Northern Electric Company Limited went on strike and picketed the premises of Nedco Ltd. In an action to restrain the Northern Electric employees from picketing Nedco, the court had to decide whether to consider Northern Electric Company Limited and Nedco Ltd. as separate corporate entities. In concluding its judgment, and stressing the exceptional nature of the facts of the case, the court stated, in part:

> After reviewing the foregoing, and many other cases, the only conclusion I can reach is this: while the principle laid down in *Salomon* v. *A. Salomon & Co., Ltd.,* supra, is and continues to be a fundamental feature of Canadian law, there are instances in which the Court can and should lift the corporate veil, but whether it does so depends upon the facts in each particular case. Moreover, the fact that the Court does lift the corporate veil for a specific purpose in no way destroys the recognition of the corporation as an independent and autonomous entity for all other purposes.
>
> In the present case Nedco Ltd. is a wholly-owned subsidiary of Northern Electric Company Limited. It was organized and incorporated to take over what was formerly a division of

[2]   [1972] 3 O.R. 829
[3]   43 D.L.R. (3d) 714

Northern Electric Company Limited. As such wholly-owned subsidiary, it is controlled, directed and dominated by Northern Electric Company Limited. Thus, viewing it from a realistic standpoint, rather than its legal form, I am of the opinion that it constitutes an integral component of Northern Electric Company Limited in the carrying on of its business. That being so, I can see no grounds upon which lawful picketing of Nedco Ltd., pursuant to a lawful strike against Northern Electric Company Limited, should be restrained.

I want to make it perfectly clear that, in reaching this conclusion, I have not attempted to lay down any general principle. It is only because of the special circumstances prevalent in this case that I have reached the conclusion which I have. While Nedco Ltd. is, for the purposes of this application, an integral component of Northern Electric Company Limited, for all other purposes it remains an autonomous and independent entity.

DURATION OF PARTNERSHIPS AND CORPORATIONS

Unless otherwise provided by the terms of a partnership agreement, pursuant to The Partnership Act of Ontario and similar statutes in other common-law provinces, a partnership is dissolved by the death or bankruptcy or insolvency of one of its partners. A corporation, on the other hand, has perpetual existence as long as the corporation complies with its governing statute, and as long as no procedural steps are taken to dissolve the corporation. The death of a shareholder does not have the effect of dissolving a corporation.

EFFECT OF PERSONAL GUARANTEES

As noted, a shareholder is not theoretically liable for the corporation's debts. However, the limited-liability characteristic of a corporation is often substantially reduced or nonexistent, as a practical matter. For example, when setting up the banking arrangements for the corporation, the incorporator is often required to sign a personal guarantee in return for satisfactory credit terms. The incorporator shareholder who signs a guarantee becomes personally obligated to the lending institution for the debts of the corporation to the extent of the guarantee. When a loan is guaranteed, the creditor is said to have "recourse" to the guarantor. The guarantor shareholder therefore loses the advantage of the limited-liability concept of the corporation's indebtedness in relation to that creditor.

BASIC TAX CONSIDERATIONS

The basic combined federal-provincial corporate income tax rate is approximately 45 percent. The rate will vary between provinces or territories, and is reduced for a corporation's manufacturing and processing income. A "Canadian-controlled private corporation" ("CCPC") for purposes of the Income Tax Act (Canada) will generally be eligible to receive a credit in respect of its tax otherwise payable on income from an active business carried on in Canada. Commonly referred to as the "small business deduction," this credit will reduce a CCPC's combined federal-provincial income tax rate on its first $200,000 per year of income from an active business to between approximately 18 and 22 percent, depending on the province or territory. The rules relating to the small business deduction are complex. For example, in the case of certain related corporations, the credit must be shared by the related group on the first $200,000 of group active business income. In addition, certain claw-back provisions apply at both the federal and provincial levels to limit or eliminate the benefit of the small business deduction in a year for CCPCs that have significant taxable capital or income in the year.

When the corporation distributes its after-tax income to its shareholder or shareholders by way of dividends, each shareholder who is an individual must pay a tax on such dividend income, and is entitled to a dividend tax credit.

In some cases, the effect of the small-business deduction is that the aggregate of the tax paid by the corporation entitled to such deduction on the income earned by it and the tax paid by the individual shareholder receiving the dividend is less than the tax that would have been paid had the business been carried on through a sole proprietorship. In other words, dividend income from a corporation can result in less tax payable than does income derived from a sole proprietorship or partnership.

In addition, there is a timing advantage available. The taxes paid by shareholders on corporate dividends are payable only when dividends are paid by the corporation. If the board of directors of the corporation chooses to defer the payment of dividends to a subsequent taxation year, then the tax payable on that dividend is deferred.

An in-depth examination of tax law is beyond the scope of this text. However, the engineer should appreciate the need for specialized tax advice in business planning.

SUMMATION OF EXCEPTIONS TO SALOMON PRINCIPLE

"Associating" corporations controlled by the same person or group of persons for tax purposes is an example of the dilution of the concept of separate and distinct corporate entities dictated by the economic realities of the business world. There are several other examples of such departures from the general concept of the distinctiveness of the corporate entity: the willingness of the courts to "pierce the corporate veil" in exceptional circumstances; the courts' disregard for the distinction between the individual and the corporate entity where fraud has been involved; the courts' "association" of corporations for certain tax purposes. Nevertheless, it is important to bear in mind that such departures are the exceptions.

THE ENGINEERING CORPORATION

The Professional Engineers Act of Ontario[4] and similar statutes governing engineering in the other common-law provinces recognize that engineers may incorporate and carry on the business of engineering as a corporation. Incorporation may provide both limited liability and tax advantages.

THE PARTNERSHIP AGREEMENT

If a decision is made to enter into a partnership, it is important to define the basis of that partnership and to execute a partnership agreement. Because of the very personal nature of the obligations that partnership creates, it is advisable to retain legal counsel for the preparation of the partnership agreement. Indeed, each partner should ideally obtain independent legal advice about the partnership agreement. Important aspects of the agreement will include: a description of the management responsibilities of each of the partners; the basis for calculating each partner's share of

[4]   R.S.O. 1990, c. P.-28

the profits or losses and contributions to working capital; provisions for dissolution of the partnership; and the basis for the withdrawal or expulsion of partners.

Partnership agreements are usually between individuals. But organizations, such as corporations, may enter into a partnership. Partnerships of corporations are not uncommon business vehicles today. When corporate partners enter into a partnership, each corporate partner's assets are at risk. The scope of each proposed partnership agreement should be closely examined to determine if the purpose of the partnership justifies that risk.

LIMITING PARTNERSHIP LIABILITY

Most of the common-law provinces have passed statutes that allow a partner to limit his or her liability. For example, The Limited Partnerships Act of Ontario[5] provides for the formation of limited partnerships, which consist of one or more general partners and one or more limited partners. Like the partners in an ordinary partnership, general partners in a limited partnership remain responsible for the debts of the firm. On the other hand, the limited partner's liability is normally limited to the amount the limited partner has contributed or has agreed to contribute. The Limited Partnerships Act of Ontario requires each limited partnership to file a certificate disclosing basic information about the partnership. The information includes the names of all general and limited partners and the amount of capital that each limited partner has contributed. A limited partner should ensure that his or her name is not used in the name of the partnership. If it is used, section 6(2) of the Limited Partnerships Act states:

> ... the limited partner is liable as a general partner to any creditor of the limited partnership who has extended credit without actual knowledge that the limited partner is not a general partner.

In effect, the limited partner may be responsible, along with the general partners, for a debt that exceeds his or her contribution to the firm.

Only general partners are authorized to transact business on behalf of the limited partnership. Pursuant to section 12(2) of The Limited Partnerships Act of Ontario, a limited partner may

[5] R.S.O. 1990, c. L.-16

from time to time examine into the state and progress of the partnership business and may advise as to its management. But the limited partner must be cautious, and have limited involvement. If the individual takes part in the control of the business, he or she can become liable as a general partner pursuant to section 13(1) of that Act.

INCORPORATION

Corporations can be formed in several ways. They can be created by statute of the federal or provincial legislatures, as are Crown corporations. More commonly, they are formed in accordance with either the Canada Business Corporations Act,[6] the Business Corporations Act of Ontario,[7] or similar statutes that govern incorporation in the other common-law provinces. Distinctions between incorporating procedures under the various statutes are not particularly important for the purpose of this text: the end result, incorporation, is essentially the same.

Both federal and provincial corporations have the capacity to carry on business beyond the geographic limits of their jurisdictions of incorporation.

In deciding whether to incorporate federally or provincially, there are certain considerations that should be borne in mind. For example, if the incorporators propose to carry on business in all provinces of Canada then federal incorporation may be appropriate. However, if the proposed business is to be carried on in Ontario and a limited number of other provinces, incorporation in Ontario may be advisable.

Provincially incorporated businesses generally require extraprovincial licences in order to carry on business in another province. A special reciprocal arrangement exists between Ontario and Quebec, which entitles businesses incorporated in either province to do business in the other without obtaining an extra-provincial licence.

OBJECTS

Neither the Canada Business Corporations Act nor the Business Corporations Act of Ontario require corporations to define their

[6] R.S.C. 1985, c. C-44
[7] R.S.O. 1990, c. B.-16

business purpose or "objects." However, each Act does permit a corporation to limit its objects, should it wish to do so. Under each statute, a corporation has the capacity and the rights, powers, and privileges of a "natural person."

All corporations that are incorporated for the purpose of carrying on the business of engineering must comply with the applicable provincial statute governing engineering.

"PRIVATE" AND "PUBLIC" CORPORATIONS

A distinction is made between "private," or closely held, corporations and "public" corporations, the shares of which are offered and distributed to the public in accordance with securities legislation and stock-exchange requirements. A "private" company is generally defined as a company in which:

(i) the right to transfer shares is restricted. (For example, such transfer may be subject to the approval of its board of directors);

(ii) the number of its shareholders, exclusive of present and former employees, is not more than fifty; and

(iii) any invitation to the public to subscribe for its securities is prohibited.

Most engineering corporations begin as private or closely held companies. A corporation might decide to "go public" and to thereby distribute its securities to the public. Such a decision will necessitate continuing compliance with extensive disclosure and reporting requirements of provincial securities legislation.

SHAREHOLDERS, DIRECTORS, AND OFFICERS

The shareholders are the "owners" of the corporation. They receive share certificates as evidence of such ownership, usually in return for invested capital.

As its owners, the shareholders elect the directors of the corporation. The board of directors of the corporation supervises the management of the corporation's affairs and business.

The officers of a corporation are elected or appointed by its directors. The officers of the corporation usually provide for the day-to-day business management. The duties of particular officers are normally set out in the by-laws of the corporation.

SHAREHOLDERS' AGREEMENTS

It is advisable for the shareholders of a closely held corporation to enter into a shareholders' agreement. An agreement commonly covers such matters as who is entitled to nominate members of the board of directors of the company; the obligations of the shareholders with respect to guarantees of the company's indebtedness; and the basis upon which issued shares of the company may be sold by a shareholder. It may also contain agreements not to communicate trade secrets of the company; or provisions to ensure that future share issuances do not dilute the respective percentage holdings of the company's shareholders.

The importance of a shareholders' agreement can be illustrated by considering the potential consequences of three individuals incorporating a company. Assume each individual takes one-third of the issued shares of the company without entering into a shareholders' agreement. Now suppose there is a "falling-out" between the parties; suppose also that two of the shareholders join forces. The third shareholder may be unable to elect a representative to the board of directors of the company. That shareholder may also be ousted from a former position as an officer and might lose his or her status as an employee of the company. The board of directors controls the declaration of dividends by the company. In our example, the board may choose not to declare dividends. Hence, the minority shareholder may be left with very little to show for a one-third shareholder interest, and might be unable to dispose of such shares. The shareholder in our example might be able to get some legal help — remedies may be available to dissenting shareholders, particularly where a "fraud on the minority" has been committed. However, it is preferable for shareholders to protect their respective interests by entering into an appropriate shareholders' agreement.

THE DIRECTOR'S STANDARD OF CARE

Directors and officers are expected to comply with a certain standard of care in carrying out their respective responsibilities. For example, Section 134(1) of the Business Corporations Act of Ontario provides:

> Every director and officer of a corporation in exercising his or her powers and discharging his or her duties shall,

(a) act honestly and in good faith with a view to the best interests of the corporation; and

(b) exercise the care, diligence and skill that a reasonably prudent person would exercise in comparable circumstances.

Any individual who consents to act as a director of a corporation must take such responsibilities seriously. An engineer who agrees to act as a director of a corporation engaged in the business of engineering must realize that the position of director has potential liabilities. The engineer must be willing to act in good faith and in the best interests of the corporation.

In addition, there are a number of statutory provisions imposing responsibilities on directors of corporations. To illustrate:

1. Pursuant to section 131 of the Business Corporations Act of Ontario, a director of a corporation is potentially personally liable for up to six months' unpaid wages of employees of the corporation, provided action is commenced against the corporation and the director in accordance with section 131.

131.(1) *Directors' liability to employees for wages* — The directors of a corporation are jointly and severally liable to the employees of the corporation for all debts not exceeding six months' wages that become payable while they are directors for services performed for the corporation and for the vacation pay accrued while they are directors for not more than twelve months under The Employment Standards Act, and the regulations thereunder, or under any collective agreement made by the corporation.

(2) Limitation — A director is liable under subsection (1) only if,

(a) the director is sued while he or she is a director or within six months after ceasing to be a director; and

(b) the action against the director is commenced within six months after the debts became payable, and

(i) the corporation is sued in the action against the director and execution against the corporation is returned unsatisfied in whole or in part, or

(ii) before or after the action is commenced the corporation goes into liquidation, is ordered to be wound up or makes an authorized assignment under the Bankruptcy Act (Canada), or a receiving order under the

Bankruptcy Act (Canada) is made against it, and in any such case, the claim for the debts is proved.

2. Section 242 of the Income Tax Act (Canada)[8] provides that directors who personally participate in the commission of offences against that Act are personally liable together with the corporation.

> 242. Where a corporation commits an offence under this Act, an officer, director or agent of the corporation who directed, authorized, assented to, acquiesced in, or participated in, the commission of the offence is a party to and guilty of the offence and is liable on conviction to the punishment provided for the offence whether or not the corporation has been prosecuted or convicted.

3. Subsection 65(3) of the Competition Act (Canada) is also relevant. It provides that, if a corporation does not properly submit certain returns, which can be required in connection with an enquiry under the Act, any director or officer of that corporation who assents to or acquiesces in the offence committed by the corporation in not filing the returns is guilty of that offence personally. The penalty for each offence is a fine of not more than $5,000, or imprisonment for not more than two years, or both.

4. Pursuant to the Corporations Information Act of Ontario[9] a corporation may not carry on business in Ontario — or identify itself to the public in Ontario — by a name or style other than its own, unless the assumed name and style are first registered under the Act. Corporations that do use assumed names must include their full corporate names on all contracts, invoices, negotiable instruments and orders for goods and services issued or made by them or on their behalf. Contravention of the Act or its regulations is an offence. If a corporation is guilty of such an offence, every director or officer of the corporation who authorized, permitted, or acquiesced in the offence is also guilty of an offence, and is liable to a fine of not more than $2,000.

5. Section 250(1) of the Canada Business Corporations Act is also relevant. It is an offence to file a report, return, notice, or other document required by the Act or its regulations that con-

---

[8] R.S.C. 1985, 5th Supplement
[9] R.S.O. 1990, c. C-39

tains an untrue statement. As well, documents cannot omit a material fact required by the Act. Where an offence is committed by a corporation, any director or officer of the corporation who knowingly authorized, permitted, or acquiesced in the offence is liable to a fine of up to $5,000, or to imprisonment for a term of up to six months, or both.

DISCLOSURE OF CONFLICTS

As noted, a director must act in good faith and in the best interests of the corporation. As well, each director is required, by statute governing the corporation, to disclose any personal interest in any material contract or transaction to which the corporation is a party. The director must not vote in approval of any such contract or transaction. If the director does not disclose interest in the contract, he or she is potentially accountable to the corporation or to its shareholders for any profit or gain realized from the contract or transaction.

THE JOINT VENTURE

The joint venture as a form of business organization has become increasingly popular amongst contractors, engineers, and architects in connection with large-scale projects, where it makes sense to join forces. A joint venture is often essentially a partnership limited to one particular project; and joint venturers should ensure that the scope of the joint venture is limited to the single project, in order to protect the assets of the joint venturers as partners in the project. It is also advisable for each of the joint venturers to indemnify each of the other joint venturers for liabilities that may arise as a result of respective services and contract obligations negligently performed. The joint-venture agreement should include a clear definition of the scope of the venture; it should also define obligations of the parties to the agreement, and the manner in which revenues and costs are to be shared.

SAMPLE CASE STUDY

The following hypothetical case and commentary is included for illustrative study purposes.

Smith is a 25 percent shareholder and director of Skylift Inc., a company engaged in commercial helicopter services in Ontario.

A friend of Smith, J. Johnson, sought Smith's technical and financial support in forming another commercial helicopter business in British Columbia and Smith agreed to so participate and acquired a 50 percent shareholder interest in the second company, known as Johnson's Skyhooks Limited.

Eventually Skylift Inc. became interested in purchasing all of the assets of Johnson's Skyhooks Limited. Smith was in no way involved in promoting the purchase of Johnson's Skyhooks Limited until the proposed purchase was presented to the five member board of directors of Skylift Inc. for approval. At that meeting, Smith did not disclose Smith's shareholder interest in Johnson's Skyhooks Limited and Smith cast the deciding vote in passing the directors' resolution to authorize the asset purchase. Shortly after the asset purchase had been finalized, the board of directors of Skylift Inc. became aware of Smith's shareholder interest in Johnson's Skyhooks Limited and, on further investigation, concluded that the price paid for the assets of Johnson's Skyhooks Limited was unreasonably high.

What action might the board of directors and shareholders of Skylift Inc. take in the circumstances? State, with reasons, the likely outcome of the action.

Commentary: In answering, reference should be made to the director's duties to act in good faith with a view to the best interests of the corporation; to the requirement to disclose conflicts; and to the consequences of not doing so; as described in Chapter 2.

CHAPTER THREE

INTERNATIONAL CONSIDERATIONS

INTRODUCTION

Foreign markets are offering attractive opportunities for Canadian engineers. The scope of the new world marketplace is vast. The burgeoning project opportunities in China, India, Malaysia, Indonesia, Vietnam, and elsewhere in Southeast Asia, as well as South America and Eastern Europe are among the destinations of opportunity and challenge.

In both developing and developed countries lack of government capital to respond to pressing infrastructure needs has resulted in privatization opportunities for the private sector. These opportunities have arisen at a time when the North American development industries are in need of new markets. Accordingly, increasing numbers of Canadian engineers are responding.

Not surprisingly, new foreign markets generate new risks. This chapter will highlight some of the important considerations to be borne in mind by Canadian engineers supplying services or products to the international marketplace.

BUSINESS ORGANIZATIONS IN FOREIGN JURISDICTIONS

The choice of business organization plays an important role in other countries just as it does in Canada. However, business organizations may be characterized somewhat differently in other jurisdictions, and tax and other issues will vary substantially from country to country. Accordingly, it is extremely important to obtain advice from an appropriately experienced lawyer in the foreign jurisdiction or to have the advice of a consultant familiar with the country. Venturing into a foreign jurisdiction is a significant step and one that shouldn't be undertaken without appropriate advice. When embarking on a business or project initiative in Southeast Asia for example, it is important to commit carefully selected senior personnel and sufficient resources to the initiative. It typically takes substantial amounts of time and investment to understand the foreign market. Critical to success in that process is understanding the foreign culture and building personal relationships with business contacts within the country. Accordingly, the Canadian engineer who embarks on such an initiative should realistically schedule the time commitment for success — typically, that time commitment will have to be a relatively lengthy one.

POLITICAL RISKS

Political risk is perhaps one of the more obvious risk factors, particularly in developing jurisdictions and countries with a history of instability. Changes in government can lead to significant policy changes that may affect business initiatives of foreign investors; changes in senior officials and bureaucrats with whom foreign investors may be dealing; and changes in local labour rates that may have a profound impact on anticipated profitability levels.

Depending upon the project and the foreign jurisdiction, some political risk insurance coverage may be available through funding agencies such as the World Bank. Accordingly, political risk insurance may provide some level of protection, but careful judgment together with prudent selection of a local influential partner to assist in assessing and managing the risks involved are extremely important factors.

FOREIGN LEGAL SYSTEMS

The basis for enforcement of rights in foreign countries may be substantially different than in Canada. Countries such as India, that have a history of colonial ties to the British legal system, offer more similarities to the Canadian legal system than others. Eastern European and other former communist countries, the Peoples' Republic of China, and Vietnam are examples of countries where the legal system is undergoing significant development. This occurrence results from internal changes and a growing interest in providing opportunities for foreign investors.

However, change and development take time and foreign investors can expect to encounter significant differences. In the area of property rights, for example, the typical North American approach to mortgage security on borrowed funds may simply not be available as part of the foreign legal system. Private property ownership rights have not been part of the communist system. In addition, the court systems in many foreign jurisdictions may well be fundamentally different from those with which Canadian engineers are familiar. These are further reasons that point to the advisability of close alliances with carefully selected local partners or joint venturers who can assist and advise with respect to compliance with local laws and customs.

Engineers are often involved in complicated projects and transactions. An example is a typical co-generation project. Canadian engineers who have participated in co-generation projects are aware of the extent and complexity of the contractual arrangements that are necessary to implement the facility. Contractual arrangements typically include gas supply contracts, gas transportation contracts, design and construction contracts, equipment purchase agreements, steam sales and electricity sales agreements, operating and maintenance agreements, loan agreements, and compliance with requirements of regulatory authorities. As complicated a process as that is in Canada, becoming involved in a co-generation project in China or elsewhere in Southeast Asia obviously requires local contacts and insights into the energy industry, not to mention appropriate contracting expertise in these jurisdictions.

Consideration of other issues relating to contracting in foreign jurisdictions is further addressed in subsequent chapters in this text, including Risks in Construction (Chapter 25); Bonds and International Performance Guarantees (Chapter 26); ADR on International Projects (Chapter 29); and the North American Free Trade Agreement (Chapter 35).

LICENSING REQUIREMENTS

Compliance with licensing requirements and obtaining necessary permits and approvals in a foreign country are important considerations that can be time-consuming. This is particularly the case when dealing with countries with a history of excessive bureaucratic procedures or inexperience in dealing with foreign investors on new types of project initiatives. Here again, the importance of local advice and relationships is emphasized. North Americans undertaking foreign project initiatives need to have a realistic assessment of the length of time that may be involved in the licensing and permitting process in a foreign jurisdiction. Personal and business licences may be required to offer engineering services in the foreign country. Whatever the nature of the foreign initiative, it is advisable to investigate well in advance as it may be a time-consuming process. A realistic view of the bureaucratic process in the foreign country is a significant factor in the important risk assessment and planning process.

In the essential licensing process, a carefully chosen local advisor or local joint venture "partner" should be in a much better position to understand the realities and to deal effectively with the bureaucracy than the foreign investor, consultant, or contractor. Knowledge of, and compliance with, local laws is most important, particularly as frustration in dealing with scheduling delays may give rise to the controversial practice of paying bribes in some countries. In Chapter 23 of this text, penalties are identified under Canadian law for engaging in offences including secret commissions, bribes, and kickbacks, as are Canadian laws relating to giving gifts or conferring benefits on government employees. Similar issues arise abroad. It is a mistake to assume that the payment of bribes in another country is not contrary to the laws of that country. It is extremely important for the Canadian engineer to be aware that even in countries where this practice may be regarded, rightly or wrongly, as commonplace, it is typically illegal and, on conviction, the penalties may be extremely severe. Accordingly, it is vital to know, and to comply with, the local laws. The development of a business or project in a particular foreign country may need to be scheduled over a sufficiently long enough time period to avoid even being drawn into this dilemma and controversy. Smart scheduling is a vital factor in the risk assessment and planning process.

FINANCIAL RISKS

Many countries impose currency exchange controls or restrictions on the transfer of funds out of the country. Changes may also occur in import duties and result from local tax policy. These may all constitute significant risks related to foreign projects that need to be carefully investigated at the outset. The risk of inflation is a very important consideration, particularly on toll road, energy projects, and similar engineering initiatives undertaken by the private sector, where future revenues generated by the project are key to its financial success. Investing Canadian or American dollars at the outset on a project where future revenues in the foreign currency may be significantly impacted by inflation within that country exposes the investor to substantial risks that require careful consideration and financial planning. Contracting for future payments in a specified currency or obtaining governmental or other guarantees, if available, to mitigate or reduce the impact of potential inflation may be among advisable approaches to deal with this risk.

Many countries have special tax incentives for certain forms of foreign investment. The choice of business entity established in the foreign country can be critical to obtaining the benefit of these tax incentives. Accordingly, setting up the optimum form of business organization in the foreign country is an important consideration that will require local advice. Whether that optimum business organization is a corporation in which shares are held by the foreign investor and by a local investor in prescribed percentages, or a joint venture approved by the foreign government, or some other form of business entity is typically a very important consideration.

CONTRACT FORMS

Another general observation that may be of interest to Canadian engineers is that the contracting forms used on projects in many foreign jurisdictions are similar to their own contract forms. For example, the "FIDIC" contract forms of the Federation Internationale des Ingenieurs Conseils, which are favoured for use on many projects financed by the World Bank, feature a third party "consultant" engineer authorized to make decisions similar to the approach contemplated by the Canadian "CCDC" contract forms. Countries such as China and Vietnam are, to some extent, basing contract approaches on "western" forms and are taking advice on such forms from commonwealth countries.

Clearly, differences in approaches to contract documents need to be understood and appreciated. Again, local assistance and expertise is advisable, particularly in understanding foreign cultures, foreign laws, and differences in approaches to negotiating agreements in the other country.

CHAPTER FOUR

TORT LIABILITY

The term "tort" has become more familiar to the engineering community as a result of the increasing frequency of claims against professionals. But its meaning may not be readily appreciated. Legal writers have not yet achieved agreement on a satisfactory precise definition of tort. The term generally refers to a private or civil wrong or injury, one that involves negligence and that may arise independently of contract. Torts are best understood by looking at some examples, and by examining the principles the courts apply to determine if tort liability exists.

Tort liability may, for example, arise from automobile accidents, from the transportation of hazardous cargoes, from the sale of unsafe products, and from the negligent performance of professional services. No privity of contract is required for tort liability to exist. Obviously, no contract exists between a negligent driver and the victim of an automobile accident, or between a company transporting dangerous explosives and a victim who sustains injuries as a result of such transportation. Even services performed gratuitously — without a contract — can give rise to liability in tort if the services are performed negligently.

Whereas no contract need exist for tort liability to arise, it is possible that tort liability and liability for breach of contract can both occur, depending on the circumstances and the terms of the contract. This is referred to as "concurrent liability in tort and contract," a concept that has now been endorsed by the Supreme Court of Canada. This concept is discussed further in Chapter 22. For the purpose of this chapter, however, the focus is on the fundamentals of tort liabilities, particularly the important emphasis placed on the engineer's "duty of care" and the measurement of what constitutes "reasonable care" in the provision of engineering services. In subsequent chapters, contract law issues relating to contracts involving a duty of care, similarly measured, will also be addressed.

FUNDAMENTAL PURPOSE

The fundamental purpose of tort law is to compensate victims of torts. Punishment of negligent wrongdoers is not a purpose of tort law. If the circumstances of the tort also constitute criminal activity, punishment of the criminal would be governed by the Criminal Code of Canada.[1] Criminal proceedings are independent of civil proceedings.

In order to ensure that funds are available to provide compensation to tort victims, engineers involved in providing design services to the public should obtain appropriate professional liability insurance coverage. (In Ontario, engineers involved in providing design services to the public must obtain appropriate professional liability insurance coverage or disclose to the client that the engineer does not have such insurance, and the client must acknowledge such disclosure.) Professional liability insurance should provide protection if an engineer's negligence results in damage arising in tort. For engineers engaged as employees in the manufacturing industries, product liability insurance may provide appropriate protection. Advice should be taken on the appropriate form of insurance coverage from experienced risk managers and insurance brokers. Some companies, government departments, and crown agencies choose to "self-insure," another approach that can suffice provided adequate funds are available to respond to compensate tort victims.

In considering insurance or other means of funding valid claims that may arise, engineers should closely examine the nature of their practices and assess what protection is most advisable. However, insurance coverage is in fact not available to cover all aspects of engineering. Exclusions in insurance policies and lack of insurance availability, for example, for pollution and nuclear hazards, should be carefully factored into the engineer's risk assessment and practice planning.

PRINCIPLES OF TORT LAW

The essential principles applicable in a tort action can be isolated to provide a basis, or formula approach, for analyzing whether tort liability arises in a given situation. In order to satisfy the court that compensation should be made, the plaintiff in a tort action must substantiate that:

[1]   R.S.C. 1985, c. C-46

(a) the defendant owed the plaintiff a duty of care,

(b) the defendant breached that duty by his or her conduct; and

(c) the defendant's conduct caused the injury to the plaintiff.

If any one of these three essential aspects of a tort action is not substantiated to the satisfaction of the court, the plaintiff will not succeed.

When tort principles are applied to a particular situation, reasonableness plays a major role.

For example: for tort liability to be established to the satisfaction of the court in a negligence action, the court must be persuaded that three things are true. First, that the defendant owed the plaintiff a duty to use a reasonable degree of care. "Reasonable" is measured by the conduct expected of a reasonable person in the circumstances. Second, that the defendant ought, reasonably, to have foreseen that failure to exercise a reasonable degree of care would likely result in injury or damage to the plaintiff. Third, that, as a result of fault on the part of the defendant (and not because of some other intervening act by a third party) the plaintiff sustained the injury. Note that reasonableness may also be a factor in determining the third item. Was the fault of the defendant the reasonable proximate cause of the damage? Or was the damage claimed too remote or not reasonably foreseeable to the defendant in the circumstances?

THE ENGINEER'S STANDARD OF CARE

A significant factor in a tort action is the establishment of the standard of care required of the defendant. For example, suppose a court is to determine whether an engineer has been negligent in the performance of engineering services. The court must apply some standard to determine whether the engineer's conduct was negligent. The standard applied is based on the premise that engineers have a duty to use the reasonable care and skill of engineers of ordinary competence. The "reasonable care" is measured by applicable professional standards of the engineering profession at the time the services are performed. In 1974, the High Court of Justice of Ontario decided *Dominion Chain Co. Ltd.* v. *Eastern Construction Co. Ltd. et al.*[2] The case concerned an action for damages arising out of the alleged faulty construction

[2] (1974) 3 O.R. (2d) 481

of a large factory roof. The court made the following statements (and references to Halsbury's text) relating to the duty of the engineer:

> Liability of engineer
>
> It is trite law that an engineer is liable for incompetence, carelessness or negligence which results in damages to his employer and he is in the same position as any other professional or skilled person who undertakes his professional work for reward and is therefore responsible if he does or omits to do his professional undertakings with an ordinary and reasonable degree of care and skill.

In 3 Hals., 3rd ed., p. 528, para. 1050, it is stated:

> Architects and engineers are bound to possess a reasonable amount of skill in the art or profession they exercise for reward, and to use a reasonable amount of care and diligence in the carrying out of work which they undertake, including the preparation of plans and specifications. Every person who enters into a profession undertakes to bring to the exercise of it a reasonable degree of care and skill, and represents himself as understanding the subject and qualified to act in the business in which he professes to act. The employer buys both skill and judgment, and the architect ought not to undertake the work if it cannot succeed, and he should know whether it will or not.

And in para. 1051 of the same volume of Halsbury, it is stated:

> As to the amount of skill required, the architect or engineer need not necessarily exercise an extra ordinary degree of skill. It is not enough to make him responsible that others of greater experience or ability might have used a greater degree of skill, or even that he might have used some greater degree. The question is whether there has been such a want of competent care and skill, leading to the bad result, as to amount to negligence.

And in paragraph 1056, pp. 530–1, it is stated:

> In addition to this, if the negligence or want of skill of the architect or engineer has occasioned loss to his employer, he will be liable to the latter in damages. These are not limited to the amount of remuneration which under the agreement the architect or engineer was to receive, but are measured by the actual loss occasioned....

The terms "negligence" and "mistake" are not necessarily synonymous. *Ramsay and Penno* v. *The King*[3] was an action against the Crown involving flooding of lands, alleging negligence in the design and construction of certain dams. The court stated, in part:

> Whether or not there was negligence in regard to design and construction of the dam is a question of fact. Engineers are expected to be possessed of reasonably competent skill in the exercise of their particular calling, but not infallible, nor is perfection expected, and the most that can be required of them is the exercise of reasonable care and prudence in the light of scientific knowledge at the time, of which they should be aware....

Accordingly, there may be circumstances where the court can be persuaded that the error in question did not amount to negligence. Such situations are likely to be rare. Arguments in support of that position might involve reference to less sophisticated technologies at the time the services were performed, thereby supporting the argument that the engineer did all he or she ought to do at that point in time. This approach is not as far-fetched as it might first appear, given the potential for claims in tort to be brought against the engineer many years in the future (as will be addressed in the next chapter on limitation periods). The argument may also have merit in some "developing" industries, perhaps aerospace or other "high tech" fields where design standards may be less precisely defined than in other long established structural and other engineering disciplines.

Section 72 of The Regulations under The Professional Engineers Act of Ontario defines "negligence" as an act or omission in the carrying out of the work of a practitioner that constitutes a failure to maintain the standards that a reasonable and prudent practitioner would maintain in the circumstances. Section 72 also codifies, by listing, a variety of specific examples of what constitutes "professional misconduct" by an engineer, and includes negligence in this list. When it comes to substantiating negligence in a lawsuit in tort, the court will look to the particular circumstances and to expert testimony to determine if there was in fact a breach of the duty of care. The description of negligence and the definitions of professional misconduct in The Regulations under

---

[3] [1952] 2 D.L.R. 819

The Professional Engineers Act of Ontario may assist the court in determining if the duty of care was breached. However, that issue will typically be settled on the basis of persuasive expert testimony by an engineer or engineers appropriately qualified.

DEVELOPMENT OF TORT LAW

There have been many significant tort case decisions. Two of the most significant cases to date have been the 1932 decision in *Donoghue* v. *Stevenson,*[4] and the 1963 decision in *Hedley Byrne & Co. Ltd.* v. *Heller & Partners Ltd.*[5] Both cases were decided by England's highest court, the House of Lords.

Donoghue v. *Stevenson* was a very important decision in the evolving field of products liability. The plaintiff had become ill after consuming the contents of a bottle of ginger beer, which had been given to the plaintiff by a friend. The bottle of ginger beer reportedly contained a decomposed snail. The House of Lords determined that the manufacturer was under a legal duty to the ultimate consumer to take reasonable care that the ginger beer was free from any defect likely to cause injury to health. (Note that no privity of contract existed between the plaintiff-consumer and the manufacturer.)

Hedley Byrne is probably the most significant case to date, as far as professionals are concerned generally. In *Hedley Byrne*, the plaintiffs were advertising agents who asked their bankers to inquire into the credit rating of a company with which the plaintiffs had business dealings. The plaintiff's bankers then made inquiries of the defendants, who were bankers for the company about whom credit information was being sought. The defendant bankers negligently reported that the company's financial position was favourable, but expressly stipulated that such advice on credit-worthiness was "without responsibility." The plaintiff proceeded to do business with the company, relying on the advice of the bankers. As a result, the plaintiff eventually lost £17,000. The House of Lords held that, had there not been an express disclaimer of responsibility from the defendant bank, the defendant bank would have been liable to provide compensation to the plaintiff for the financial loss that resulted from the defendant bank's negligent misrepresentation. Implicit in the decision of the House of Lords was the belief that, where one person relied

---

[4]   [1932] A.C. 562
[5]   [1964] A.C. 465

on the special skill and judgment of another, and when the second person knew of that reliance, the second person was duty bound to take reasonable care in exercising the special skill.

A 1994 decision of the Ontario court clearly confirmed that statements disclaiming responsibility to third parties will absolve the party making the statements from liability, as far as third parties are concerned. The case is *Wolverine Tube (Canada) Inc. v. Noranda Metal Industries Ltd. et al.*[6] Noranda asserted a claim against an environmental consultant, alleging negligence in carrying out environmental compliance audits and liability assessments on certain properties Noranda was contemplating selling. The environmental consultant had included the following statement in each of its reports:

> This report was prepared by Arthur D. Little of Canada, Limited for the account of Noranda, Inc. The material in it reflects Arthur D. Little's best judgment in light of the information available to it at the time of preparation. Any use which a third party makes of this report, or any reliance on or decisions to be made based on it, are the responsibility of such third parties. Arthur D. Little accepts no responsibility for damages, if any, suffered by any third party as a result of decisions made or actions based on this report.

Subsequently the properties were sold to Wolverine. Wolverine had no dealings with the environmental consultant, who was unaware of the sales until five years later when the lawsuit commenced. However, Noranda advised Wolverine that it could rely on the consultant's reports and it needn't engage its own environmental consultant.

In deciding that the consultant's disclaimers should be effective to preclude liability, Justice Jennings made some interesting remarks. He stated, in part, as follows:

> It is apparent that Canadian authorities binding upon me have consistently recognized not only the duty of care for negligent misrepresentation as found in *Hedley Byrne*, but also the right of the issuer of the statement to disclaim any assumption of a duty of care....
>
> I have been invited by counsel for Wolverine to consider that the language in the statement is not sufficiently broad to insulate Little from the negligence claimed here. I do not think

[6]   21 O.R. (3d) 264

it is profitable to engage in an overly close scrutiny of the precise meaning of the words in the statement. I content myself with saying that in my opinion the language is far more comprehensive than that which insulated the defendant bank in *Hedley Byrne.*

Prior to the *Hedley Byrne* decision, tort law provided relief where damages to person or property had been incurred. *Hedley Byrne* expanded the scope of damages to include financial loss that resulted from advice negligently given — where the person giving the advice knew, or ought to have known, that reliance was being placed on his or her skill and judgment. The case is significant for two reasons. First, it expanded the scope of damages that may be recovered in a tort action. Second, it focussed attention on services performed by professionals who possessed special skills.

The principles enunciated in the *Hedley Byrne* decision have since been applied in cases involving engineers. One example is the 1979 decision of the Manitoba Queen's Bench, in the case of *Trident Construction Ltd.* v. *W.L. Wardrop and Assoc. et al.*[7] An engineer provided services on a project that involved the construction of a sewage disposal plant. The engineer was held liable to the contractor on the project because of his unsuitable design. No privity of contract existed between the contractor and the engineer. In applying the *Hedley Byrne* decision, the Manitoba court made reference to the summary given by the authors of *Charlesworth on Negligence* (5th ed., p. 32, para 49):

> The House of Lords has thus expressed the opinion that if in the ordinary course of business including professional affairs a person seeks advice or information from another who is not under any contractual or fiduciary obligation to give it, in circumstances in which a reasonable man so asked would know he was being trusted or that his skill or judgment was being relied on, and such person then chooses to give the requested advice or information without clearly disclaiming any responsibility for it, then he accepts a legal duty to exercise such care as the circumstances require in making his reply; for a failure to exercise that care, an account for negligence will lie if damage or loss results.

---

[7] [1979] 6 W.W.R. 481

Subsequently, the Manitoba court concluded:

> Why should it be otherwise, as to the responsibility of the professional engineer or architect whose plans the builder is required to follow, in the event a mistake in those plans proves costly to the builder? Surely, the party whose design it is may be taken to have in contemplation the party invited to build the project as designed, and who by his contract will have to abide by the plans in question, as forming an integral part of his undertaking with the owner. I have no difficulty in fixing the professional engineer with a duty of care towards the person who is to follow the engineer's design, to ensure that the plans are workable, for breach of which duty the engineer may be made accountable.

A similar example of the application of the *Hedley Byrne* decision is the 1983 judgment of the Ontario High Court in *Brown & Huston Ltd.* v. *The Corporation of the City of York et al.*,[8] another case in which the contractor succeeded against consulting engineers who had omitted certain important information relating to a soils report and ground-water levels. The contract involved the construction of an underground pumping station and required the contractor to satisfy himself, by a personal examination, as to the conditions he would encounter during construction. The court found that, because the engineers omitted information, the contractor bid on the job assuming there was no water problem. The court concluded that the contractor plaintiff was partly negligent because he did not ask about the soils report, and he did not learn the precise meaning of the information the engineers provided. But the court held the consulting engineers liable and apportioned responsibility for damages seventy-five percent against the engineers and twenty-five percent against the contractor. In considering the specific nature of the negligence of the consulting engineers, the court stated, in part:

> ... The engineers must have known that tenderers would rely on the tender package; particularly when the contract documents did not require the contractor to satisfy itself about the sub-surface conditions.
>
> Was the lack of reference to the soils reports and the change of the sketch and plan a negligent omission to convey necessary

---

[8]   5 C.L.R. 240

information? Information concerning the water and sub-surface conditions was of great significance to any tenderer. I can think of no good reason why the engineers did not refer to the soils reports in the tender package and no reason for this omission was advanced at trial.

I am satisfied that the plaintiff company relied on the tender package in preparing its bid and bid the job on the basis that there was not a water problem. Mr. Leonardelli saw the inverted triangle and recognized it as a water mark but thought it could well be surface water. Having concluded that there was no water problem, or more properly, not having recognized a water problem, the proposed method of excavation was reasonable. I am also satisfied that had the plaintiff recognized the water problem the bid would have been different or the plaintiff would not have made a bid at all.

The main defence is that the information given in the tender package and particularly SK-1 and SK-2 and the description of work in the Schedule of Unit Prices which required close timber sheeting should have made it clear to the plaintiff that there was a water problem. In this connection I do not accept the evidence of Mr. Mitchell that the inverted triangle would be understood by a reasonable contractor to be the stabilized ground water level. His evidence in this connection is not in accord with the evidence of Mr. Cummings and Mr. King. It is also contrary to the CSA standard (Ex. 27) above referred to.

I also accept the evidence adduced on behalf of the plaintiff that the term "close timber sheeting" from the description of work means "close timber sheeting as required". I have come to the conclusion that there was reasonable reliance on the lack of information concerning a water condition so as to bring the case within the principles enunciated in *Hedley Byrne & Co. v. Heller & Partners Ltd.*

Principles from *Hedley Byrne* were also applied at the Ontario Supreme Court level in a November, 1983 decision, *Unit Farm Concrete Products Ltd.* v. *Eckerlea Acres Ltd. et al.*; *Canama Contracting Ltd.* v. *Huffman et al.*[9] In this case, a contractor was engaged by an owner to design and construct a barn to be placed over a manure pit. The contractor succeeded in bringing an action against an engineer who was employed by the Department of Agriculture of Ontario. The court decided that the plaintiff

[9]   5 C.L.R. 149

contractor was entitled to rely on the advice of the defendant engineer, a government employee and not a consulting engineer. From time to time over the years, the plaintiff contractor had relied upon advice given by the defendant engineer, who was employed by the government to assist farmers and contractors to work out their plans. The judgment also indicates that the contractor and the engineer had never met to discuss the plans, but had discussed the matter by telephone. A copy of the plans had been left on the engineer's desk by the contractor. The engineer acknowledged at trial that he did not carefully review the plans. However, he did look through the plans and he forwarded the following written message to the contractor:

> Good set of plans. I like the detail. Wish I could spend that amount of time on each project. Keep up the good work.

The court noted that the plans had two particular deficiencies. One: the plans showed the reinforcing rod to be in the middle of the wall. The rebar should have been closer to the inside of the wall for maximum support. Two: there was a complete absence of any rebar schedule on the plans. Hence the structural-steel components and requirements vital to the integrity of the cement wall were missing. Expert evidence established, to the court's satisfaction, that it would be part of the duty of the engineer to note the deficiencies in the plans. The court concluded that the engineer failed in his duty by giving wrong advice with respect to the rebar schedule and by not reading the plans carefully and not noting the deficiencies. Also, as the court noted, the engineer was negligent in referring to the plans as "good plans," the effect of which "would be to lull the plaintiff into thinking the plans were adequate." As a result of the deficiencies, portions of the manure-tank walls failed.

The defendant engineer argued that he did not know he was being consulted. The court pointed out that, when "being held to account for negligence, it is not what we subjectively feel or think but what our conduct objectively makes the other person believe we feel or think."

In the Canama case, the court was also critical of the performance of the contractor. The court concluded that the engineer had failed to comply to the standard of care of a reasonable engineer, but also concluded that the contractor had failed to comply to the standard of care of a reasonable designer and builder in the circumstances. The court particularly noted the manner in which the contractor placed the plans on the engineer's desk, pointing out that the contractor had a duty, as a designer, to make it clear to the engineer that the engineer was being asked for advice. The contractor failed in that duty. The contractor and the engineer were each found to be fifty percent at fault in causing the damage to the walls of the manure tank.

In the Canama case, the court also pointed out that, had the defendant engineer clearly disclaimed responsibility for his words, as occurred in *Hedley Byrne*, the engineer could have avoided responsibility. Although the contractor was partly responsible, the court judged for the plaintiff against the defendant engineer and his employer, Her Majesty The Queen, because the defendant engineer was a crown servant and an employee of the Province of Ontario at the time.

The Canama case was appealed by the contractor to the Ontario Court of Appeal. The Court of Appeal noted the circumstances in which the engineer had provided the advice, but nevertheless increased the engineer's liability, reapportioning responsibility and holding the engineer seventy-five percent, and the contractor twenty-five percent, responsible for the damage.

An illustrative tort case involving mechanical engineers is *SEDCO and Hospital Laundry Services of Regina* v. *William Kelly Holdings Ltd. et al.*,[10] a 1988 Saskatchewan Queen's Bench decision. The owner SEDCO leased a newly constructed building to Hospital Laundry Services. SEDCO had contracted with an architect for the building design, who in turn contracted with the defendant mechanical engineers. The mechanical design was faulty, resulting in significant mechanical defects, primarily in relation to the building's cooling system. Hospital Laundry Services succeeded in its tort claim in its capacity as a tenant against the subconsultant mechanical engineers. It claimed for losses due to employees having to take too frequent "heat breaks." The court emphasized that the engineers had known from the outset that the building was to be custom-designed for one tenant, Hospital Laundry Services, and that they knew that worker comfort was a major consideration. Furthermore, they knew or ought to have known that a faulty design for a cooling system would lead to high temperatures in the building with a corresponding drop in worker productivity and a resulting financial loss for Hospital Laundry Services. The court stated that the purely economic dam-

---

[10] [1988] 4 W.W.R. 221 (Sask. QB)

ages were recoverable, especially in circumstances where the breach of the duty of care constituted a foreseeable danger to the health or safety of an individual.

An important 1993 decision of the Supreme Court of Canada confirmed that an engineering firm preparing drawings and specifications can be liable in tort to a contractor with whom the engineering firm has no contractual relationship. The case is *Edgeworth Construction Ltd. v. N.D. Lea & Associates Ltd.* The British Columbia Court of Appeal had held that the engineering firm owed no duty of care to the contractor. The Supreme Court of Canada disagreed and overruled the decision of the British Columbia court. In deciding that the engineering firm could be liable in tort to the contractor who lost money on the project due to errors in the specifications and construction drawings, the Supreme Court of Canada included the following, as part of its analysis:

> Liability for negligent misrepresentation arises where a person makes a representation knowing that another may rely on it, and the plaintiff in fact relies on the representation to its detriment: *Hedley Byrne & Co. Ltd.* v. *Heller & Partners Ltd.*, [1964] A.C. 465 (H.L.); *Haig* v. *Bamford* (1976), 72 D.L.R. (3d) 68, 27 C.P.R. (2d) 149, [1977] 1 S.C.R. 466.
>
> The facts alleged in this case meet this test, leaving the contract between the contractor and the province to one side. The engineers undertook to provide information (the tender package) for use by a definable group of persons with whom it did not have any contractual relationship. The purpose of supplying the information was to allow tenderers to prepare a price to be submitted. The engineers knew this. The plaintiff contractor was one of the tenderers. It relied on the information prepared by the engineers in preparing its bid. Its reliance upon the engineers' work was reasonable. It alleges it suffered loss as a consequence. These facts establish a *prima facie* cause of action against the engineering firm....
>
> The contract, by cl. 42, stipulated that any representations in the tender documents were "furnished merely for the general information of bidders and [were] not in anywise warranted or guaranteed by or on behalf of the Minister ...". This arguably absolved the province from any liability for the plans. The exemption extends, on its express words, only to warranties "by or on behalf of the Minister". It does not

purport to protect the engineers against liability for their representations.[11]

The availability of tort remedies on construction projects in Canada was again confirmed by the Supreme Court of Canada in early 1995. The Supreme Court of Canada in its decision in *Winnipeg Condominium Corporation No. 36* v. *Bird Construction Co. Ltd.*, in reversing a decision of the Court of Appeal for Manitoba, confirmed that contractors (as well as subcontractors, architects, and engineers) who take part in the design and construction of a building will owe a duty in tort to subsequent purchasers of the building if it can be shown that it was foreseeable that a failure to take reasonable care in constructing the building would create defects that pose a substantial danger to the health and safety of the occupants. In such circumstances, liability arises for the reasonable cost of repairing the defects and putting the building back into a non-dangerous state. In reaching its conclusion, the Supreme Court of Canada rejected the U.K. House of Lords precedent followed by the Court of Appeal for Manitoba. The *Winnipeg Condominium* case is indicative of the importance of tort remedies in Canada, notwithstanding previous decisions that have emphasized the necessity for contractual relationships in order for liability for damages to arise in the construction industry.

The *Winnipeg Condominium* case involved a building that was originally completed in 1974 and converted into a condominium in 1978. In 1982, the condominium corporation's directors became concerned about the masonry work on the exterior cladding of the building and retained consultants who offered the opinion that the building was structurally sound. In 1989, a storey-high section of the cladding, approximately 20 feet in length, fell from the ninth storey level of the building to the ground below. Removal and replacement of the cladding was carried out at a cost in excess of $1.5 million.

In his very considered review of the issues and the precedents, Justice La Forest referred to the concern in such circumstances that allowing recovery for economic loss in tort will subject a defendant to what Cardozo C.J. in *Ultramares Corp.* v. *Touche*, 174 N.E. 441 (N.Y.C.A. 1931), at p. 444, called "liability in an indeterminate amount for an indeterminate time to an indeterminate class." However, that concern was not sufficient to change

---

[11] (1994) 107 D.L.R. (4th) 173

the outcome of the *Winnipeg Condominium* decision, which affirmed the potential in Canada for liability in tort.

The Supreme Court of Canada also made note of the following excerpt from the decision of the Supreme Court of South Carolina, in *Terlinde* v. *Neely*, 271 S.E.2d 768 (1980), at p. 770:

> The key inquiry is foreseeability, not privity. In our mobile society, it is clearly foreseeable that more than the original purchaser will seek to enjoy the fruits of the builder's efforts. The plaintiffs, being a member of the class for which the home was constructed, are entitled to a duty of care in construction commensurate with industry standards. In light of the fact that the home was constructed as speculative, the home builder cannot reasonably argue he envisioned anything but a class of purchasers. By placing this product into the stream of commerce, the builder owes a duty of care to those who will use his product, so as to render him accountable for negligent workmanship.

In concluding and ordering the case to proceed to trial, even though no contractual tie existed between the original contractor and the subsequent purchaser of the building, Justice La Forest stated, in part:

> I conclude, then, that no adequate policy considerations exist to negate a contractor's duty in tort to subsequent purchasers of a building to take reasonable care in constructing the building, and to ensure that the building does not contain defects that pose foreseeable and substantial danger to the health and safety of the occupants. In my view, the Manitoba Court of Appeal erred in deciding that Bird could not, in principle, be held liable in tort to the Condominium Corporation for the reasonable cost of repairing the defects and putting the building back into a non-dangerous state. These costs are recoverable economic loss under the law of tort in Canada.

STRICT LIABILITY

The discussion of torts to this point has emphasized the concept of fault; we have concentrated on cases where the conduct of the party that caused the injury was unsatisfactory in terms of duty owed. However, our legislators have sometimes found the application of the concept of fault inadequate for the purpose of compensating injured parties. For example, workers' compensation legislation recognizes that fault is not necessary if compensation is to be provided. All employers are expected to make contributions on behalf of employees and if an employee negligently injures himself or herself, compensation is provided according to provincial workers' compensation legislation.

In products liability cases in the United States, a manufacturer may be strictly liable for any damage that results from the use of the product even though the manufacturer was not negligent in producing it. Canadian products-liability law has not yet adopted this "strict liability" concept, but the law appears to be developing in that direction.

VICARIOUS LIABILITY

Our courts have long recognized the concept that the employer is vicariously liable for the negligent performance of an employee. If an employee commits a tort in the course of employment, the employer will be vicariously liable for the damage caused. This concept may appear onerous as far as the employer is concerned, but it is consistent with the basic premise of tort law; its purpose is to compensate the injured party. The employer provides compensation because it is presumed that the employer is in a better financial position than the employee.

In 1972, England's Court of Appeal decided the case of *Dutton* v. *Bognor Regis United Building Co. Ltd.*[12] Foundations laid by the builder of a house were discovered to be inadequate to carry the load of the building, and damage resulted. The house had been built on a rubbish deposit and the foundations should have been deeper to withstand the pressure of settling. Building by-laws required that the building's foundations be approved by a local building inspector before construction continued. The inspector failed to make proper inspection before giving approval. The local building authority that employed the inspector was held liable to a subsequent purchaser of the house for the inspector's negligence. In his reasons for judgment, Lord Denning examined the question of the liability of the inspector:

> It is at this point that I must draw a distinction between the several categories of professional men. I can well see that in the case of a professional man who gives advice on financial

[12] [1972], 1 All E.R. 462

or property matters — such as a banker, a lawyer or an accountant — his duty is only to those who rely on him and suffer financial loss in consequence. But, in the case of a professional man who gives advice on the safety of buildings, or machines, or material, his duty is to all those who may suffer injury in case his advice is bad. In *Candler* v. *Crane, Christmas & Co.*, I put the case of an analyst who negligently certifies to a manufacturer of food that a particular ingredient is harmless, whereas it is in fact poisonous; or the case of an inspector of lifts who negligently reports that a particular lift is safe, whereas it is in fact dangerous. It was accepted that the analyst and the lift inspector would be liable to any person who was injured by consuming the food or using the lift. Since that case the courts have had the instance of an architect or engineer. If he designs a house or a bridge so negligently that it falls down, he is liable to everyone of those who are injured in the fall: see *Clay* v. *A. J. Crump & Sons Ltd.* None of those injured would have relied on the architect or the engineer. None of them would have known whether an architect or engineer was employed, or not. But beyond doubt, the architect and engineer would be liable. The reason is not because those injured relied on him, but because he knew, or ought to have known, that such persons might be injured if he did his work badly.

The action was framed against the local building authority and not against the inspector in his personal capacity. Nevertheless, the Court of Appeal made it clear that the inspector was one of those responsible:

First, Mrs. Dutton has suffered a grievous loss. The house fell down without any fault of hers. She is in no position herself to bear the loss. Who ought in justice to bear it? I should think those who were responsible. Who are they? In the first place, the builder was responsible. It was he who laid the foundations so badly that the house fell down. In the second place, the council's inspector was responsible. It was his job to examine the foundations to see if they would take the load of the house. He failed to do it properly. In the third place, the council should answer for his failure. They were entrusted by Parliament with the task of seeing that houses were properly built. They received public funds for the purpose. The very object was to protect purchasers and occupiers of houses. Yet, they failed to protect them. Their shoulders are broad enough to bear the loss.

Employees are also potentially liable in tort. An example is the case of *Northwestern Mutual Insurance Co.* v. *J.T. O'Bryan & Co.*,[13] decided in 1974 by the British Columbia Court of Appeal. Northwestern asked its agent to delete a certain risk from a policy. This was a standard practice in the insurance industry. The employee of the agency negligently assured Northwestern that the risk had been deleted. The company relied on that assurance. Subsequently, however, Northwestern was required to pay on the risk. On the basis of *Hedley Byrne*, the British Columbia Court of Appeal held both the agency and its employee liable for negligence. The court found that the employee owed a duty of care towards the insurance company, Northwestern, which he breached by his total lack of care. That lack of care was the sole cause of the damage.

Hence tort liability can apply vicariously to the employer, and the employee will also be personally liable for the tort the employee has committed. To protect its employee engineers, therefore, a corporation providing engineering services should ensure that its professional liability insurance policy extends to cover the liability of both the corporation and its employee engineers.

However, the 1993 *Edgeworth Construction* decision of the Supreme Court of Canada (referred to earlier in this chapter) confirms that Canadian courts may be prepared to depart from the traditional approach to employees' liability in certain circumstances where employee engineers are concerned. The Supreme Court of Canada made some very important statements about the potential liability of an engineering firm, and of individual employee engineers.

A seven-member panel of the Supreme Court of Canada unanimously decided that the contractor was entitled to sue the engineering firm but was not entitled to recover from the individual engineers.

Not all of the judges agreed precisely upon the reasoning that precluded the lawsuit against the individual engineers. Six of the seven judges found that the individuals owed no duty of care to the contractor, and therefore could not be liable to the contractor. (As indicated, there was no contract between the individual and the contractor.) They adopted the following statements:

> The only basis upon which [the individual engineers] are sued is the fact that each of them affixed his seal to the design

---

[13]   51 D.L.R. (3d) 693

documents. In my view, this is insufficient to establish a duty of care between the individual engineers and [the Contractor]. The seal attests that a qualified engineer prepared the drawing. It is not a guarantee of accuracy. The affixation of a seal, without more, is insufficient to found liability for negligent misrepresentation.

The remaining judge indicated that he was in general agreement with the others, but added some interesting comments about the individual engineers. In particular, he said:

> ... there are sound reasons of policy why [the individual engineers] should not be subjected to a duty to the [Contractor]. The [Contractor] here was quite reasonably relying on the skills of the engineering firm....
> The situation of the individual engineers is quite different. While they may, in one sense, have expected that persons in the position of the [Contractor] would rely on their work, they would expect that the [Contractor] would place reliance on their firm's pocketbook and not theirs for indemnification ....

While the reasoning of all of the judges may give some comfort to individual engineers, it does not support the conclusion that an individual engineer employed by an engineering firm will never be personally liable in tort. At least one of the judges indicates that an individual may be liable where the facts of the case are such that there was reliance on the particular individual, as opposed to that individual's employer.

CONCURRENT TORTFEASORS

At times, torts concur to produce the same damage. It is possible for more than one party to be liable in such a tort action. The defendants are said to be "concurrent tortfeasors."

An example is the 1979 decision of the British Columbia Court of Appeal in *Corporation of District of Surrey* v. *Carrol-Hatch et al.*[14] An architect had designed a new police station, and had engaged a firm of engineers to perform structural design services. The building eventually underwent "extensive structural change" because of settlement problems. The problems could have been avoided had proper soils tests been conducted. After

---

[14] [1979] 6 W.W.R. 289

examining two shallow test pits, the engineers had recommended to the architect that deep soils tests be taken. But the architect had rejected the recommendation, and the engineers had submitted a "soils report" to the owner on the basis of a superficial examination of the shallow test pits only. Both the architect and the engineers were held liable to the owner. The Appeal Court agreed with the trial judge's apportionment of fault — 60 percent to the architect and 40 percent to the engineers — in the circumstances. The court held that the architect and engineers were concurrent tortfeasors; they had breached their duty to warn the owner that additional soils tests should be taken.

PRODUCTS LIABILITY

Products liability in Canada is not yet premised on strict liability, as it is in some jurisdictions in the United States. Canadian courts continue to apply principles of negligence in products-liability matters. Generally, where the plaintiff can establish that damage has clearly resulted from appropriate use of a product, the defendant manufacturer must then persuade the court that, considering the state of the particular industry's technological advance at the time, the manufacturer could not have foreseen the defective nature of the goods manufactured. Otherwise liability will arise.

Products liability has developed through considerations of both contract law and tort principles. The tort concept of fault has been applied to extend the limits of the scope of products liability; the development has also been premised on implied contractual warranties. This is a logical overlap between contracts and torts: a contract of sale is essential, at some point, for products liability to arise. For example, The Sale of Goods Act in Ontario[15] and similar statutes in the other common-law provinces imply various conditions and warranties in a sale-of-goods contract. One condition is that the goods will be merchantable and reasonably fit for the purpose for which they are sold. (Where a vendor attempts to limit warranty obligation by contract, consumers may have the benefit of provincial consumer protection statutes requiring manufacturers to provide warranty protection to the consuming public.)

[15] R.S.O. 1990 c. S.1

An awareness of products-liability matters is important to engineers as professionals who may be engaged in manufacturing or sales, as well as to engineers as consumers.

The scope of products liability has substantially increased since the following principles were enunciated in the *Donoghue* v. *Stevenson case:*[16]

> A manufacturer of products which he sells in such a form as to show that he intends them to reach the ultimate consumer in the form in which they left him, with no reasonable possibility of intermediate examination, and with the knowledge that the absence of reasonable care in the preparation or putting up of the products will result in injury to the consumer's life or property, owes a duty to the consumer to take that reasonable care.

The *Donoghue* v. *Stevenson* case referred to the "manufacturer of products." Our courts have now extended the duty of care to others; for example, assemblers, installers, sub-manufacturers, importers, wholesalers, retailers, distributers, repairers, and business suppliers. Wherever it can be established that injury ought reasonably to have been foreseen in any particular circumstances, a potential for products liability arises.

STANDARD OF CARE AND DUTY TO WARN

Risk of injury is inherent in some products. A manufacturer must warn the consumer of any dangerous potential of the product by appropriate labelling.

For example, in 1976 the Alberta Supreme Court (Appellate Division) decided the case of *George Ho Lem* v. *Barotto Sports Ltd. and Ponsness-Warren, Inc.*[17] The plaintiff, an experienced hunter, purchased a shot-shell reloading machine that was in no way defective; if operated in accordance with its clear instructions, the machine would produce only normal shot-shells. The plaintiff received personal instruction on the use of the machine, but he did not follow instructions; nor did he follow the instruction manual. He did not realize the consequences of not following the instructions. The machine mismanufactured some shot-shells, the chamber of the plaintiff's gun burst on firing, and the plaintiff was injured.

---

[16] Supra, at page 36
[17] (1976) 1 C.C.L.T. 180

The shot-shell reloading machine was manufactured by one defendant and sold to the plaintiff by another defendant. The plaintiff's claim was that the defendants had failed in their duty to warn him adequately of the possibility of mismanufacture of a shot-shell, which would nevertheless be normal in appearance.

The Appellate Court held that adequate instructions for the use of the machine had, in fact, been given. The plaintiff lost his case because he had not followed clear and simple instructions. The manufacturer's responsibility was to warn of dangers related to its product; this warning was given. The damage suffered by the plaintiff was not caused by failure in a duty owed to him by either of the defendants. Rather, the court held that damage was caused by the plaintiff's own fault. The manufacturer had met the very high standard of care expected of manufacturers of potentially dangerous products.

Not all manufacturers succeed in meeting that high standard of care, in the court's opinion. An example is the 1971 decision of the Supreme Court of Canada in *Lambert* v. *Lastoplex Chemicals Co. Limited et al.*[18] The male plaintiff was a consulting engineer who had graduated in mechanical engineering. He and his wife, the co-plaintiff, jointly owned a home, to which the plaintiff was doing some repairs. He purchased two one-gallon cans of a fast-drying lacquer sealer manufactured by one of the defendants. The plaintiff proposed to use it to seal a parquet floor, which he was installing in the recreation room of his home. The recreation room was located in the basement of the house; it was separated from the furnace and utility room by a plywood wall and by a fireplace. There was a door opening at the northerly end between the two rooms, but there was no door. In the furnace and utility room there was a natural-gas furnace and a natural-gas water heater, both of which had pilot lights.

The cans of lacquer sealer bore certain caution notices on the labels. The plaintiff read the labels before starting to apply the lacquer sealer. However, during the application, one or both of the pilot lights in the furnace and utility room came in contact with the fumes or vapours of the lacquer sealer. There was an explosion and consequent damage when fire reached one of the half-full cans of lacquer sealer, which was open.

The product containers bore three separate warnings that the product was inflammable. But such warnings were inadequate, in the court's view. As stated by the court:

[18] [1972] S.C.R. 569

The three labels on the cans of the respondent's product contained, respectively, the following cautions: (1) The largest label, rectangular in shape, which bore the name and description of the product, contained on its end panel, in addition to drying time information, the words "Caution inflammable! Keep away from open flame!" Along the side of this panel vertically and in small type, were the words "Danger — harmful if swallowed, avoid prolonged skin contact, use with adequate ventilation, keep out of reach of children". (2) A diamond-shaped red label with black lettering, issued in conformity with packing and marking regulations of the then Board of Transport Commissioners for Canada and having shipping in view, had on it in large letters the following: "KEEP AWAY FROM FIRE, HEAT AND OPEN-FLAME LIGHTS", "CAUTION", "LEAKING Packages Must be Removed to a Safe Place", "DO NOT DROP". (3) A third label, rectangular in shape, contained a four-language caution, which was in the following English version: "CAUTION, INFLAMMABLE — Do not use near open flame or while smoking. Ventilate room while using".

The evidence disclosed that a lacquer sealer sold by a competitor of the respondent contained on its label a more explicit warning of danger in the following terms: "DANGER — FLAMMABLE," "DO NOT SMOKE. ADEQUATE VENTILATION TO THE OUTSIDE MUST BE PROVIDED. ALL SPARK PRODUCING DEVICES AND OPEN FLAMES (FURNACES, ALL PILOT LIGHTS, SPARK-PRODUCING SWITCHES, ETC.) MUST BE ELIMINATED, IN OR NEAR WORKING AREA."

On appeal, the Supreme Court of Canada found in favour of the plaintiff. According to the court, in labelling the product, the manufacturer had failed to warn a reasonable user of the danger of the pilot lights. The case emphasizes the need for extreme caution on the part of manufacturers in labelling products.

In its decision, the Supreme Court of Canada also made the following statements:

Manufacturers owe a duty to consumers of their products to see that there are no defects in manufacture which are likely to give rise to injury in the ordinary course of use. Their duty does not, however, end if the product, although suitable for the purpose for which it is manufactured and marketed, is at the same time dangerous to use; and if they are aware of its dangerous character they cannot, without more, pass the risk of injury to the consumer.

The applicable principle of law according to which the positions of the parties in this case should be assessed may be stated as follows. Where manufactured products are put on the market for ultimate purchase and use by the general public and carry danger (in this case, by reason of high inflammability), although put to the use for which they are intended, the manufacturer, knowing of their hazardous nature, has a duty to specify the attendant dangers, which it must be taken to appreciate in a detail not known to the ordinary consumer or user. A general warning, as for example, that the product is inflammable, will not suffice where the likelihood of fire may be increased according to the surroundings in which it may reasonably be expected that the product will be used. The required explicitness of the warning will, of course, vary with the danger likely to be encountered in the ordinary use of the product.

In the *Lambert* v. *Lastoplex Chemicals* case, Justice Laskin made an additional comment that points out the high standard of care imposed on the manufacturer and may be of interest to the engineer-consumer. The manufacturer was unable to avoid liability although he emphasized that the injured party was a qualified engineer:

The question of special knowledge of the male appellant was argued in this Court as going to the duty of the respondent to him and not to his contributory negligence. What was relied on by the respondent as special knowledge was the fact that the male appellant had qualified as a professional engineer, he knew from his experience that a lacquer sealer was inflammable and gave off vapours, and hence knew that it was dangerous to work with the product near a flame. This, however, does not go far enough to warrant a conclusion that the respondent, having regard to the cautions on the labels, had discharged its duty to the male appellant.

ECONOMIC LOSS

In the *Hedley Byrne*[19] decision, a tort matter, economic losses resulted from advice negligently given. In products-liability mat-

---

[19] Supra at page 36

ters, however, there was a reluctance to extend liability for negligence to economic losses in the absence of actual physical injury until the 1973 decision of the Supreme Court of Canada in *Rivtow Marine Ltd.* v. *Washington Iron Works et al.*[20] The court decided that economic losses caused by the use of a defective product may, in some circumstances, be recoverable. In the *Rivtow Marine* case, the plaintiff chartered a logging barge, which was fitted with a crane manufactured by one defendant — Washington Iron Works, an American corporation — and distributed in Canada by another defendant. Washington Iron Works had also manufactured a second crane, virtually identical to the one on the logging barge chartered by the plaintiff. The second crane had been installed on a similar barge and had collapsed. The crane operator was killed, and there was an investigation by the then Workmen's Compensation Board of British Columbia. Very serious structural defects were found in the crane chartered by the plaintiff. The defects were similar to those that were later found to have caused the death of the crane operator. It was established, from the evidence, that the defendants had both been aware for some time that the cranes were subject to cracking as a result of negligence in design. But neither of the defendants had taken steps to warn the plaintiff of the potential danger and necessity for repair. The Supreme Court of Canada held that the defendants were under a duty to warn the plaintiff of the necessity for repairs as soon as they became aware of the defects and the potential danger. Thus the manufacturers were liable to the plaintiff in negligence for the economic loss attributable to their failure to warn. In other words, they were liable for lost profits while the crane was out of service for repairs.

The potential for liability for economic loss in products-liability matters was established in the *Rivtow Marine* decision. In 1977 the British Columbia Supreme Court decided the case of *MacMillan Bloedel Ltd.* v. *Foundation Co.*[21] In the *MacMillan Bloedel* case, the defendant's workers negligently damaged an electric cable that supplied electricity to the office building of MacMillan Bloedel. This damage interrupted the supply of electricity. As a result, the plaintiff was unable to continue its operations and was forced to send its employees home for the day. The

---

[20] (1974) 40 D.L.R. (3d) 530
[21] (1977) 1 C.C.L.T. 358

plaintiff sought to recover the amount of the salaries and wages paid to its staff ($48,841.00). The court concluded that the defendant ought to have foreseen the economic harm caused to the plaintiff. The loss was a direct result of the defendant's negligence — the damage to the electrical cable. The court indicated that if economic loss was suffered by the plaintiff as a result of the defendant's negligent act, such loss was not too remote to be compensated. However, the court was not satisfied that the employees' salaries constituted an economic loss that resulted from the defendant's negligence. As the court pointed out, the salary payments were payable to the employees in any event. The court had no other evidence before it of economic loss that resulted from negligence, so the plaintiff's action was dismissed. But the court did indicate that it was willing to award damages for economic loss, if such economic loss could be properly accounted for.

However, the courts have imposed limitations on when economic loss may be compensated. In 1977, the Federal Court, Trial Division, decided the case of *Bethlehem Steel Corporation* v. *St. Lawrence Seaway Authority*.[22] The court asserted that, where there has been no damage to person or property in which the claimant might have some interest, the right to recovery for pure economic loss remains very limited.

In the *Bethlehem Steel Corporation* case, a ship ran into a lift bridge over a canal, destroying the bridge and obstructing the canal. As a result, shipping through the canal was delayed for several days. The court found that the owner of the ship that struck the bridge was legally responsible. When the time came to distribute funds that had been paid into court, the validity of two particular claims became an issue. One of the claimants asked for loss of profits for two of its ships, which had been delayed for about two weeks. The other claimant asked for the cost of shipping certain cargo overland to Toronto, where it could be loaded for shipment to Europe. Neither claim was allowed.

The court stated that the *Rivtow Marine* case did not change the law. At best, the court suggested the *Rivtow Marine* case only extended liability to economic loss where there had been physical harm to the property of the claimant, or where such physical harm had been threatened. As there was no harm or threat of harm to the claimants' property in the *Bethlehem Steel* case, recovery for purely economic loss was denied.

[22] 79 D.L.R. (3d) 522

However, a 1982 decision of the English House of Lords in *Junior Books Ltd.* v. *Veitchi Co. Ltd.*[23] may well influence future Canadian court decisions. The decision actually extended liability for consequential economic loss to a negligent subcontractor. A negligently laid floor was defective but had not caused danger to the health or safety of any person, or risk of damage to any other property belonging to the owner. The House of Lords pointed out that the defendant subcontractor was a specialist whose skill and knowledge was relied upon. No direct contractual relationship existed between the owner and the subcontractor, but the House of Lords noted that the relationship fell only just short of a direct contractual relationship. The damage caused to the owner was a direct and foreseeable result of the subcontractor's negligence in laying the defective floor. The subcontractors were in breach of a duty owed to take reasonable care. The subcontractor was held liable for the foreseeable consequences of the breach, including such items as the costs of replacing the floor, storage of books during the carrying out of remedial work, and removal of machinery to enable the remedial work to be carried out; lost profits due to business disturbance; lost wages; and overhead costs.

In 1990, the Supreme Court of Canada referred to the *Junior Books* decision in a case involving a claim for economic loss. The case was *Canadian National Railway Co.* v. *Norsk Pacific Steamship Co.*[24] A log barge had negligently collided with a bridge owned by Public Works Canada, which was used by C.N.R. to cross the Fraser River. The Supreme Court of Canada was satisfied that there was sufficient "proximity" between the plaintiff and the defendants to justify liability. Although owned by Public Works Canada, the tracks to and from the bridge were the property of C.N.R. The court held that the captain of the tugboat ought to have foreseen the economic loss that would result from the damage negligently caused, given the necessity for C.N.R. to reroute traffic during repairs. C.N.R. was so closely linked to Public Works Canada in the circumstances that it was within the reasonable ambit of risk. Given the "sufficient proximity," C.N.R. was entitled to recover its economic loss.

The 1995 decision of the Supreme Court of Canada in the *Winnipeg Condominium* case, as previously referred to, further

[23] 20 [1982] 3 All ER 201
[24] 65 D.L.R. (4th) 321

confirms that economic loss is recoverable, in that case for the costs of repairing defects and correcting a dangerous cladding condition. The key element emphasized by the Supreme Court in *Winnipeg Condominium* was foreseeability.

OTHER RELEVANT TORTS

There are many different classifications of torts. Some of the other torts that may be relevant to engineers include:

(1) the tort of defamation, which is further divided into two classifications: libel and slander. In essence, the reputation of the plaintiff is damaged by untrue statements publicly made by the defendant. If the statements are made in writing, the tort is referred to as "libel"; if the statements are verbal, the tort is referred to as "slander." If statements that damage a reputation are true, no liability arises.

(2) occupiers' liability. The occupier of property must exercise the required standard of care to ensure the safety of individuals coming onto that property. A duty of care extends to trespassers, although trespassers are not accorded a standard of care as are those coming onto the property for business reasons or as guests. Guests must be safeguarded against dangers the occupier is aware of. A higher duty is owed to those who come onto the property for business reasons; business visitors must be safeguarded against damages the occupier is aware of or ought to be aware of as a reasonable person. The occupier is under an obligation not to deliberately harm a trespasser; for example, he cannot set traps. In Ontario, The Occupiers' Liability Act[25] supersedes the common law. Liability is now governed by the Act, which specifies duties of care generally consistent with common-law principles. But the Act does not recognize the previous general common-law distinction between business and social guests.

(3) the tort of nuisance, designed to alleviate undue interference with the comfortable and convenient enjoyment of the plaintiff's land. Nuisance, as a tort, is growing in potential application, particularly as environmental issues have become

---

[25] R.S.O. 1990, c. 0.2

extremely high profile. For example, in 1972, the British Columbia Supreme Court decided *Newman et al.* v. *Conair Aviation Ltd. et al.*[26] The defendant aviation company's insecticide spray drifted onto lands for which the spray was not intended, and nominal damages were awarded. Another example is the 1974 decision of the Ontario Court of Appeal in *Jackson et al.* v. *Drury Construction Co. Ltd.*[27] Blasting operations by a contractor resulted in fissures opening up in the granite bedrock. The fissures allowed material in a barnyard to escape into the percolating waters feeding the plaintiff's well. The Court of Appeal concluded that the contractor should be liable. The following excerpt from the judgment of the Court of Appeal is of interest:

It is to be observed that a blasting operation such as this cannot be viewed as a natural use of the land. It is inherently a dangerous operation. As long as the percolating waters found their way to the plaintiffs' premises, the plaintiffs ought not to have been deprived of its beneficial properties. Surely if, in the course of excavation, deleterious materials flow from the equipment onto the plaintiffs' lands or are placed directly on the plaintiffs' premises as a result of the dynamiting, it can scarcely be argued that the plaintiffs would not have a cause of action. In my view, the same principle should govern where the plaintiffs' percolating waters are polluted as a direct result of the defendant's blasting operations, even though the pollution comes from a source other than the defendant's property or premises.

SAMPLE CASE STUDIES

The following hypothetical cases and sample answer (or commentaries) are included for illustrative study purposes:

1. A manufacturing company retained an architect to design a new plant. The manufacturer, as client, and the architect entered into a written client/architect agreement in connection with the project. The purpose of the plant construction was to enable the client to expand its manufacturing and warehousing facilities.

The structural design of the plant was prepared by an engineering firm that was retained by the architect. A separate

[26] 33 D.L.R. (3d) 475
[27] 4 O.R. (2d) 735

agreement was entered into between the architect and the engineering firm to which the client was not a party.

The engineering firm turned the matter over to one of its employees, a professional engineer with experience in structural steel design who proceeded to complete the structural design of the plant. The client had informed the architect that the second floor of the plant was to be used for manufacturing and warehousing purposes and that forklift trucks would be extensively used in both the manufacturing and warehousing sections on the second floor. The architect passed this information on to the engineering firm. The employee engineer designed a steel frame and specified that the second floor was to be a concrete-steel composite, consisting essentially of concrete poured onto a steel deck, and containing a light steel mesh. The steel deck, concrete thickness, and steel mesh specifications were detailed in the engineer's design and were taken from design tables from the engineering firm's library that had been published by a company which manufactured and supplied the steel deck.

The construction of the plant was completed and, shortly after manufacturing commenced at the plant, severe cracks appeared in the concrete on the second floor. After two months of operation the floor cracked and broke up so badly that the plant had to be shut down and a remedial floor slab, heavily reinforced with reinforcing bar, was poured on top of the damaged second floor.

The design of the remedial floor slab was carried out by another consulting engineering firm. After completing its investigation of the cause of the failure of the second floor, the second engineering firm stated that, in its opinion, the engineer who had designed the second floor had used design tables from the steel deck manufacturer that were 12 years out of date and had also failed to use the tables that the engineer obviously ought to have used knowing that the floor was intended for manufacturing and forklift truck loading. The second consulting engineering firm concluded that the depth of concrete and size of steel mesh in the floor as initially designed resulted in a floor that might have been appropriate for the design of an office or apartment building but not for manufacturing and warehousing purposes.

What potential liabilities in *tort law* arise from the preceding set of facts? In your answer, state the essential principles applicable in a tort action and apply these principles to the facts. Indicate a likely outcome of the matter.

Sample Answer:

1. The fundamental purpose of tort law is to compensate a party that has suffered damages as a result of a negligent act or omission. The potential for tort liability can arise even where there is no contractual relationship between the plaintiff and the defendant.

In order for a claim in tort to succeed, the plaintiff in a tort action must prove on a "balance of probabilities" that:

(a) the defendant owed the plaintiff a duty of care;

(b) the defendant breached that duty by his or her conduct; and

(c) the defendant's breach caused the injury to the plaintiff.

As this is a tort question, it is unnecessary to address actions for breach of contract for the purposes of this answer (even though claims in contract would also be brought as a practical matter). On the facts, the logical plaintiff would be the manufacturer and the likely defendant the engineer and the engineering firm.

The engineer's duty of care was to perform its services using a reasonable degree of care and skill. The engineer knew, as an employee of the engineering firm retained to prepare the structural design, that the manufacturer would be reasonably relying on the sufficiency of the design, and the engineer therefore had a duty to take reasonable care. The engineer ought to have reasonably foreseen that reliance would be placed on the engineer's design. The engineer ought also to have reasonably foreseen the damages that the engineer could cause to the manufacturer by virtue of any negligence on the engineer's part.

Expert engineering testimony could be brought, on the facts, to substantiate that the engineer failed to use up-to-date design tables as the engineer ought to have done. (This is how the court will measure the engineer's performance against the objective standard to be applied of what constituted a reasonable degree of care and skill.) Further, the engineer ought to have taken into account the conditions to which the floor would be subjected and to have reasonably foreseen the consequences of not doing so. On the facts, the engineer breached the engineer's duty to take reasonable care.

The final prerequisite for tort liability to arise is that damages must actually have been incurred. Here the damages included the necessary costs associated with repairing the cracked floor and the possible economic losses due to the plant shutdown.

Given that the engineer was negligent and the essential tort law elements have been satisfied, the engineer's employer is automatically vicariously liable for engineer's tort. Holding the engineering firm vicariously liable is consistent with the fundamental purpose of tort law, which is to compensate the injured party, insofar as the employer is assumed to be in a better financial position to pay for the damages. This is sometimes referred to as the "deep pockets theory."

2. An engineer may be asked to provide engineering services to friends, neighbours, or community organizations of which the engineer may be a member. Can engineers ever be liable for losses or damages arising when engineering services are provided without any fee being charged? Explain.

Commentary: In answering, remember that no contractual relationship is necessary for tort liability to arise. Accordingly, a fee is not necessary for tort liability to arise. Explain the prerequisites in tort and how they would be applied.

3. Acme Manufacturing Limited designed and manufactured two identical cranes, which were sold by an Ottawa dealer to two companies contracting logging barge services on the Ottawa River. The names of the purchasers were Movemore Limited and Unjammers Inc. Both cranes went into service at approximately the same time. After one month's service the crane purchased by Movemore Limited collapsed, killing the operator of Movemore's barge.

During the Workers' Compensation Board's investigation into the accident, it became apparent that the cause of the collapse was the negligent structural design of the crane. It also became apparent that the manufacturer and the dealer had been aware of the structural weaknesses for some time. Three weeks later, upon learning that a crane of similar design had been sold to Unjammers Inc., representatives of the Workers' Compensation Board notified Unjammers Inc. of the disaster involving the sister crane sold to Movemore Limited. At no time had Acme Manufacturing Limited or the Ottawa dealer notified Movemore Limited or Unjammers Inc. of any potential problems.

As a result, Unjammers Inc., at the height of its busiest season, was forced to return the crane to the factory for repairs.

What claim can Unjammers Inc. make against the crane manufacturer or against the dealer in the circumstances? Explain.

Commentary: This is very similar to the *Rivtow* case. Explain the importance of the duty to "warn" in your answer and the potential for an economic loss claim in answering this tort question.

CHAPTER FIVE

LIMITATION PERIODS

Limitations statutes of the common-law provinces generally provide that tort actions and actions for breach of contract must be commenced within prescribed time periods after the time the cause of action "arose." An action may be commenced at any time within the prescribed period. Should the action be commenced at a later date, it will fail. For tort actions in Ontario, the prescribed period generally is six years from the "time the cause of action arose." In contract, unless the contract expressly limits the time period (as is often the case), the prescribed period is generally six years but is extended to twenty years where the contract is signed under seal. An action commencing after the prescribed period is said to be "statute barred." There are, however, a number of statutes that expressly provide for other limitation periods, and such statutes will govern where applicable. Note that Section 46 of the Professional Engineers Act of Ontario provides that an action against a member, licensee, or holder of a certificate of authorization for damages that arise when the member provides a service within the practice of professional engineering must be commenced no later than twelve months after the service was, or ought to have been, performed. The court can, however, extend the time limitation if it is satisfied there are reasonable grounds.

In 1980, the Ontario Divisional Court considered the time limitation in *Attorney-General of Canada* v. *Libling et al.*[1] The court pointed out that it had to consider all the circumstances of the particular case, including the reason for the delay; the extent to which the plaintiff acted properly and reasonably once he knew whether the act or omission of the defendant engineer might give rise to a lawsuit; the increase in difficulty of proof that

[1] 29 O.R. (2d) 44

accompanies the passage of time; and any prejudice to either the plaintiff or the defendant engineer if the extension of the limitation period was or was not granted. In the *Libling* case, the defendant engineers said another engineer was responsible for certain roof-design problems. The court noted that attempts had been made to correct the problem with the roof, but the engineer in question had not been told of the problem for almost eleven years. He had no records of the roof design. The court was not prepared to extend the limitation period.

However, engineers should appreciate that the courts are empowered by the Professional Engineers Act to extend the limitation period. They may do so if circumstances are appropriate.

The Professional Engineers Act of Ontario does not use the more general reference found in limitation statutes: "from the time the cause of action arose." Instead, the Act refers, as indicated, to the date of performance of services. But contractors generally still need to understand the phrase "from the time the cause of action arose." The "time the cause of action arose" might not be when the contractor performs the work; it might be many years later, when deficiencies that resulted from the contractor's negligence start to appear. And the claim might conceivably be made in tort by someone who was not even a party to the original contract — a subsequent purchaser of a house, for example. Thus the contractor's potential liability may be extended for a remarkable period of time. In circumstances when both contractor and engineer are responsible for any deficiencies, a court might extend the limitation period against the engineer pursuant to the Professional Engineers Act of Ontario. The question of extended limitation periods is a matter of considerable controversy amongst contractors and professionals engaged in the construction industry. A review of certain case decisions is important to an understanding of the law.

The English courts have accepted the principle that the limitation period starts only when the damage manifests itself and the plaintiff first discovers it, or ought, with reasonable diligence, to have discovered it. An example is the 1976 English Court of Appeal decision in *Sparham Souter et al.* v. *Town & Country Developments (Essex) Ltd. et al.*[2] The case involved negligence in the erection of a building. The Court of Appeal considered the question of the time when the cause arises and stated:

[2]   (1976) C.A. 858

The cause of action accrues when the damage caused by the negligent act is suffered by the plaintiff and that cannot be before that damage is first detected, or could by the exercise of reasonable skill or diligence have been detected.

In 1976, the Ontario Court of Appeal heard the case of *Dominion Chain Ltd.*[3] The trial judge found that a roof failure was caused by the negligent construction procedures of the contractor and by the negligent performance of professional skills by the engineering firm on the project. The Court of Appeal considered the question of when a cause of action arises in tort and referred to the *Sparham Souter* decision as establishing the legal principle; that the cause of action arises when the damage is first detected, or ought to be detected. The reference by the Ontario Court of Appeal to the *Sparham Souter* decision represented a departure: in previous Canadian case decisions, the "time the cause of action arises" was held to be when the services were negligently performed.[4] It therefore became arguable that the principle enunciated in the *Sparham Souter* decision was applicable in Canada; the design professional could conceivably be liable for negligent design services performed more than six years before the defect is discovered.

The Ontario High Court of Justice, however, rejected the application of the *Sparham Souter* decision in its 1981 decision in *Robert Simpson Co. Ltd.* v. *Foundation Co.*[5] The case involved alleged negligence in the design, manufacture, and installation of certain ceiling anchors, as well as alleged negligent misrepresentation that materials and method of construction were adequate. Justice Holland stated, in part:

> In the present case I am of the view that the damage occurred when the inadequate anchors were incorporated in the building, but, in any event, no later than the date at which the building was turned over to Simpsons. In my opinion the law of Ontario has not yet adopted the test which found favour in *Sparham Souter* v. *Town and Country Developments (Essex) Ltd....*

As more than six years had elapsed since the damage had so occurred, the Ontario High Court held that the limitation period had expired and the plaintiff could not succeed.

[3] 68 D.L.R. (3d) 385
[4] Schwebel v. Telekes (1967) 1 O.R. 541
[5] (1981) 34 O.R. 2d (1)

However, the decision of the Ontario High Court in the *Robert Simpson* case was reversed on appeal. In its 1981 decision in the case,[6] the Ontario Court of Appeal applied the concept endorsed by the English courts: in tort, the cause of action against the contractor arose not necessarily when the work was performed, but rather at such time as the damage was first detected or ought reasonably to have been detected.

In fact, in another case in 1981, the Ontario High Court of Justice reached the same conclusion. In *Viscount Machine and Tool Ltd.* v. *Clarke*,[7] Justice Henry held that the six-year limitation period during which an action in tort could be commenced against a negligent land surveyor commenced not when the negligent act was done but when the damage was discovered, more than six years after the survey had been negligently performed.

The decision of the Ontario Court of Appeal in the *Robert Simpson* case confirmed the application of the English precedent, the *Sparham-Souter* case, but the matter was further complicated by a subsequent English House of Lords decision. In 1982, the House of Lords heard *Pirelli General Cable Works Ltd.* v. *Oscar Faber and Partners*.[8] In the *Pirelli* case, the House of Lords expressly overruled the decision of the English Court of Appeal in *Sparham-Souter*, holding that the limitation period commences when physical damage occurred to the building whether or not the damage could have been detected by the exercise of reasonable skill or diligence at the date the physical damage occurred. The *Pirelli* case involved the construction of a chimney about one hundred sixty feet high. The chimney was made of precast concrete. Unfortunately, the concrete used for the refractory inner lining was partly made of a relatively new material, called Lytag, which was unsuitable for the purpose. Cracks developed, and eventually the chimney had to be taken down and replaced. The House of Lords acknowledged that it was difficult to determine when the damage actually occurred. But they pointed out that that is, in any event, a matter of evidence. In *Pirelli*, the House of Lords indicated that new legislation would be desirable, to "remedy the unsatisfactory state of the law on this subject." In overruling the *Sparham-Souter* decision, the House pointed out that the concept endorsed in *Sparham-Souter* — of a "date of discoverability" — could lead to the investigation of

---

[6] (1982) 360 O.R. (2d) 97
[7] 126 D.L.R. (3d) 160
[8] [1983] 1 All ER 65

facts many years after their occurrence. This might be unfair to a defendant unless a "final longstop date" could be prescribed, perhaps by statute.

However, the Supreme Court of Canada considered and decided not to adopt the *Pirelli* case in a significant 1984 decision in *City of Kamloops* v. *Nielsen et al.*[9] The Supreme Court confirmed that the municipality was liable to a house purchaser for the negligence of its building inspector, who failed to enforce a stop-work order during construction. The inspector had issued the stop-work order because the foundations had not been constructed in accordance with the approved plans. At the time of construction, the owner did not want to comply with the stop-work order. The construction was completed without complying with the order; eventually, the foundations subsided (the trial judge found the owner seventy-five percent responsible and the municipality twenty-five percent responsible for causing damages). No occupancy permit was issued. However, a plumbing permit was issued about eight months after the stop-work order was issued. About two years after the house was finished, it was sold to the plaintiff, who discovered that the foundations had subsided. The municipality was held to be liable for twenty-five percent of the "economic loss" to the plaintiff. The "loss" was the cost of reconstruction plus ancillary expenses (the Supreme Court of Canada distinguishing *Kamloops* v. *Nielson* from the *Rivtow Marine* case).

In deciding not to follow the *Pirelli* case but to confirm the application, in Canada, of the *Sparham-Souter* principle in a tort action — that the limitation period commences when the damage is first discovered, or ought with reasonable diligence to have been discovered by the plaintiff — the Supreme Court concluded:

> There are obvious problems in applying Pirelli. To what extent does physical damage have to have manifested itself? Is a hair-line crack enough or does there have to be a more substantial manifestation? And what of an owner who discovers that his building is constructed of materials which will cause it to collapse in 5 years' time? According to Pirelli he has no cause of action until it starts to crumble. But perhaps the most serious concern is the injustice of a law which statute-bars a claim before the plaintiff is even aware of its existence....

---

[9] [1984] 29 C.C.L.T. 97

Lord Fraser and Lord Scarman were clearly concerned over this.... The only solution in their eyes was the intervention of the Legislature.

This Court is in the happy position of being free to adopt or reject Pirelli. I would reject it. This is not to say that Sparham-Souter, supra, presents no problem. As Lord Fraser pointed out in Pirelli the postponement of the accrual of the cause of action until the date of discoverability may involve the Courts in the investigation of facts many years after their occurrence. Dennis v. Charnwood Borough Council, [1982] 3 All E.R. 486 (C.A.) is a classic illustration of this. It seems to me, however, to be much the lesser of two evils.

As already noted, limitation periods may vary significantly depending upon the nature of any claim or lawsuit and depending also against whom the action is to be commenced. Accordingly, as limitation periods may be relatively short (as in construction liens and other matters such as actions against public authorities and professional engineers, in some cases), it is important that legal advice with respect to the limitation period relevant to any claim be obtained as soon as possible, if there is any doubt.

The "traditional" approach of our courts to determining the limitation period during which a breach of contract action can be commenced is to measure the applicable limitation period (6 years for simple contracts; 20 years for contracts under seal; unless another limitation period is specified in the contract) from when the contract was breached.

However, this "traditional" approach was eroded by the 1985 decision of the Ontario Court of Appeal in *Consumers Glass* v. *Foundation Co. of Canada*,[10] in which the Court of Appeal held, in effect, that in cases based on a breach of a duty of care, whether in contract or tort, in the absence of an express contractual limitation period the limitation period does not begin until the plaintiff discovers or ought reasonably to have discovered the damage. *Consumers Glass* broadens the scope of the application of the "discoverability" concept from tort cases to contract cases where a duty of care is involved.

The *Consumers Glass* case involved both design and construction contract obligations relating to a warehouse constructed in 1963 and a roof collapse in 1981. There was no contract provision addressing the limitation period issue (no contract docu-

[10] (1985), 51 O.R. (2d) 385

ments could be located). The defendants took the position that the relevant limitation period in contract had long since expired and that the action was therefore statute barred. Acknowledging the right to sue concurrently in tort and contract, the Ontario Court of Appeal rejected the defendants' argument and held that the "discoverability" concept applied, notwithstanding the contractual relationships.

Although the principle in the *Consumers Glass* case has not been tested before the Supreme Court of Canada, the decision clearly points to the advisability of appropriate contractual protection as far as limitation periods in tort and contract are concerned.

SAMPLE CASE STUDIES

The following hypothetical cases and commentaries are included for illustrative study purposes.

1. The question of how long an engineer or a contractor can be sued for negligence or breach of contract is one that is of concern to professional engineers and to contractors. Describe the limitation periods during which engineers and contractors can be sued in tort or contract.

1. *Commentary*: Answer should list the limitation periods: 6 years from "breach" in simple contract (20 years under seal); 6 years from discoverability in tort; 12 months for engineers in Ontario, subject to extension. Mention the possibility of applying "discoverability" concept in the contract if no limitation period is included in the contract.

2. A contractor, in designing and constructing a shopping centre in 1983, negligently designed and installed certain ceiling anchors. The shopping centre was sold to a new owner in 1990. The inadequacy of the ceiling anchors was not detected until September of 1991 when the new owner undertook significant renovations and discovered that new ceiling anchors and new ceilings had to be installed. Is the new owner entitled to recover any damages from the contractor? In your answer, describe the limitation periods during which engineers and contractors can be sued in tort.

2. *Commentary*: Point out that tort liability does not require privity of contract; describe tort limitation periods and point out different limitation periods in tort for engineers and contractors.

CHAPTER SIX

PROOF

THE BURDEN OF PROOF

In civil proceedings, such as actions in tort or contract, the plaintiff must generally prove the case against the defendant by persuading the court on a "balance of probabilities" that the facts are as the plaintiff alleges them, and that the defendant should be held liable. In certain criminal proceedings, an accused person must be proven guilty "beyond a reasonable doubt"; proof "on the balance of probabilities" is obviously different. Much has been written about these two degrees of proof. The degree of proof implied by the term "on the balance of probabilities" is obviously less than the degree of proving guilt "beyond a reasonable doubt."

In hearing a particular case, a court may be faced with conflicting testimony from witnesses; the court must decide which of the witnesses is more credible. A judge may have to accept one person's word rather than another's. As difficult as that may seem, it is not an unusual experience for a judge.

ENGINEERS AS EXPERT WITNESSES

Engineers often find themselves making appearances as expert witnesses in court. As an expert witness, the engineer often plays a vital and persuasive role. The expert is permitted to express opinions with respect to his or her area of expertise; the witness should be cautious, and restrict testimony to such areas. Non-expert witnesses are not usually permitted to express opinions; they are restricted to establishing the facts of the case.

The expert witness is usually enlisted by one of the disputants, and is thus allied with one "side" of a case. An engineer who acts as an expert witness must take these duties seriously. He or she can expect to be subjected to cross-examination by counsel for the other party. The engineer should not undertake to appear as an expert witness unless confident of handling cross-examination. It is likely that counsel for the other side will attempt, during cross-examination, to dissuade the court from accepting the engineer's opinions.

Preparation is of the utmost importance in litigation. The expert witness should clearly understand the issues in the lawsuit and be aware of the scope of questions that can be reasonably expected.

CHAPTER SEVEN

CONTRACTS

It is important for the engineer in business to understand the essential elements of a contract. For a contract to be binding and enforceable, five elements must be present:

1. an offer made and accepted;
2. mutual intent to enter into the contract;
3. consideration;
4. capacity to contract;
5. lawful purpose.

Within the framework of these essential elements and in accordance with the contract rules the courts have developed, parties choose terms and conditions to define the nature of the agreement between them. The private nature of the law of contracts thus becomes most evident.

Contracts consist of benefits to and obligations of the contracting parties. Agreements are generally arrived at by choice or through negotiation. The law will enforce the provisions of a valid contract; the law will not intervene to impose contract terms more favourable than those negotiated between the parties.

In certain circumstances the law may intervene to declare a contract void, voidable, or unenforceable; some of these circumstances will be discussed later in this text. However, the engineer must be aware of one basic premise: if a "bad business deal" is negotiated, the courts will not impose more favourable terms.

Parties can, however, always alter an existing contractual arrangement by mutual agreement, provided the amendment is effected within the framework of the essential contract elements.

ASSIGNMENT OF RIGHTS

Contractual benefits (i.e., rights arising pursuant to the contract; for example, the right to receive payment for services rendered) can be assigned to a third party by one of the contracting parties without the consent of the other party to the contract. For example, book debts or accounts receivable can be assigned. If contracting parties wish to limit such assignment rights, they should expressly provide that no rights under the contract are assignable to a third party without the written consent of the other contracting party.

CHAPTER EIGHT

OFFER AND ACCEPTANCE

An offer is a promise made by one person — the offeror — to another — the offeree. It may involve a promise to supply certain goods or services on certain terms, for example.

Not all contracts must be in writing. The offer may be communicated orally. For the purpose of evidence, however, it is preferable to effect communications in writing.

Until it is accepted, the offer may be withdrawn by the offeror unless it is made expressly and effectively irrevocable by its terms. The offer will lapse if it is not accepted within a reasonable period of time.

If the offeree does not accept all the terms of the offer but purports to accept the offer subject to a variation in its terms, no contract is formed. Rather, a counter-offer has been made by the offeree, who thereby becomes the offeror.

Acceptance of an offer must be clearly communicated.

Business offers are usually made subject to express terms. For example, a business might offer to supply a specified machine at a quoted price. The offer might contain a proviso — that acceptance of the offer can be made during a limited time period. The offer might also state that acceptance must be communicated in accordance with the terms of the quotation.

IRREVOCABLE OFFERS

An offeree might want to ensure that an offer will not be revoked by the offeror before the offeree can accept it. This normally occurs in the tendering process, for example. Upon instructions from the owner, bidders submit offers or tenders that have been made irrevocable for a specific period of time. At any point during that period, the offer may be accepted and a contract will be formed. For reasons that will be noted, "contract consideration" is necessary where such irrevocable offers are submitted or such bids must be submitted under seal, in order to be binding.

THE OPTION CONTRACT

The option contract is another means of keeping an offer open for a certain period of time. The right to accept the offer is preserved until the offeree chooses to exercise the option. The offeror is thus precluded from revoking the offer. Something of value — for example, a payment of nominal amount — must be made at the time of entering into the option agreement in order to make the option contract enforceable.

An option agreement may be advantageous in many business situations. For example, an individual might want to purchase a particular business, but may be unwilling to make a firm offer until having completed a review of its financial and other business records. To prevent the owner from selling the business until completion of the investigation, the prospective purchaser may be able to persuade the owner to enter into an option agreement. For an agreed price (which is usually substantially less than the total purchase price of the business), the owner of the business becomes obligated to sell only to the prospective purchaser, during a specified period. And the owner can sell only upon the terms set out in the option agreement. A specific time limit is stipulated in the option to purchase agreement. If not exercised by the specified time, the exclusive option to purchase will expire.

Option agreements are particularly common in mining contracts. The party purchasing the option might wish to carry out exploration work before deciding to expend a large sum in acquiring property rights, for example.

Purchasing options might also be desirable in land development. A prospective purchaser might want to find out if it is possible to acquire various pieces of land from various owners in a particular area before making a large expenditure on an overall land purchase for development purposes.

MANNER OF COMMUNICATION

(a) Timing

Accepting an Offer   The law has developed certain general "rules" to specify when communications are effective. There are several different ways to communicate. For example, suppose the parties establish the mail as the means of communication between them. If one party decides to accept an offer, the acceptance is effected when posted. Another example: the parties might establish the telegram as the means of communication. The "rule": a communication is effected at the time the message is delivered to the telegraph operator.

Unless the two parties agree to communicate by post or telegram, the communication of the acceptance of an offer is effected only when it is actually received by the offeror.

Revoking an Offer   Similar rules, however, do not apply to the timing of the revocation of an offer. Notwithstanding what means of communication is used, the general rule is that revocation is not effective until the offeree actually receives notice of the revocation.

Complications can arise. For example, an offeror might decide to revoke his offer. He communicates the revocation through the mails, but his letter takes a few days to reach the offeree. In the meantime, the offeree has written, effecting the acceptance of the offer by mailing her acceptance. The acceptance is valid. Thus, any offeror who intends to revoke an offer should do so as expeditiously as possible, by telephone or telex, for example.

(b) Governing Law

There is also a "rule" that relates to the determination of the law applicable to a contract. The general rule is that the law of the place where the acceptance of the offer becomes effective is applicable (unless otherwise agreed upon). For example, suppose an equipment supplier located in the Province of Ontario has provided a quotation by mail to a prospective customer located in the State of New York. The equipment supplier should expressly state in the quotation that any contract resulting from the acceptance of the offer will be governed by the laws of the Province of Ontario. The supplier will thus avoid the argument that the law of the State of New York will apply to the contract (if, for example, acceptance of the offer is communicated by mail from New York State).

In 1983, the Nova Scotia Supreme Court, Appeal Division, decided *The Queen et al.* v. *Commercial Credit Corp. Ltd.*[1] The decision confirms the "rule" that says acceptance is effective on

---

[1]   4 D.L.R. (4th) 314

despatch when the acceptance is mailed, and confirms that the rule also applies where acceptance is made by courier. In the *Commercial Credit* case, an offer of conditional sales was prepared by Commercial Credit, who was acting as agent for the parties to the contract. The offer was sent by courier to the offeree, who was located outside Nova Scotia. The acceptance was sent back by courier. The court held that the contract was therefore not made within Nova Scotia. Therefore a particular Nova Scotia statute, the Instalment Payment Contracts Act, did not apply. In its decision, the court stated in part:

> I am inclined to agree with the conclusion of the trial judge that the mailbox doctrine does apply here, although for slightly different reasons. It seems to me that in relation to both of the conditional sale contracts in question the purchaser and the vendors had nominated Commercial Credit as their agent to arrange for the execution of the contract documents. Commercial Credit employees prepared the documents and made certain that they were properly executed and then advanced the funds. They were the ones that chose the method of communication and having done so on behalf of both parties the mailbox doctrine was brought into play. Its extension to a courier service was sound in principle and, in my opinion, the contracts were therefore made outside of Nova Scotia when their acceptances were sent back to Commercial Credit. The Instalment Payment Contracts Act does not apply.

THE BATTLE OF THE FORMS

An equipment supplier usually attaches to its quotations certain terms and conditions pursuant to which the manufacturer is prepared to sell its product. These terms and conditions may describe warranty rights, terms of payment, termination, governing law, indemnities, etc. Often the offeree purports to accept the offer; but this offer is actually a counter-offer, because the offeree demands terms and conditions of sale that differ from those attached to the original offer or quotation. Engineers involved in such equipment supply contracts should be cautious to ensure that acceptance of an offer (or counter-offer) is not made without a clear understanding of the terms and conditions that apply to the sale. As a general rule, the terms and conditions stipulated in an accepted counter-offer will prevail unless the person examining the counter-offer takes exception to any unsatisfactory terms or conditions.

Terms and conditions of sale should be examined and negotiated. As a practical matter, however, the terms and conditions — the "fine print" — are too often overlooked. If a contract dispute later occurs, the resolution of the dispute could be complicated. Sufficient attention should be paid to terms and conditions at the time of acceptance of the offer to prevent unnecessary complications.

SAMPLE CASE STUDY

The following hypothetical case and commentary is included for study purposes.

Acme Ltd., a manufacturing company, wished to expand its business by purchasing the business of one of its competitors, Jones Brothers Limited. The president of Acme met with the president of Jones Brothers to discuss the possibility of the purchase. At the conclusion of the meeting, both presidents felt that the matter was worth pursuing. The president of Acme presented a letter to the president of Jones Brothers:

Jones Brothers Limited,   December 28, 1994
Toronto, Ontario.

Attention: Harold Jones, President

Dear Sirs:

This letter will confirm our interest in purchasing, and your interest in selling, all of the assets, undertaking and business of Jones Brothers Limited for a total price of approximately $1,000,000, subject to the preparation of audited financial statements of Jones Brothers Limited and subject to our further negotiation of terms and conditions to apply to such sale, pending the finalization of which we agree to enter into a mutually satisfactory agreement to so acquire the business of Jones Brothers Limited.

If you are in agreement with the contents of this letter, would you please execute the duplicate enclosed copy and return.

Yours very truly,
Acme Ltd.
Per: Jane Finnigan, President

As requested, the president of Jones Brothers signed and returned, by personal delivery, the enclosed copy of the letter.

Several weeks went by and the auditors of Jones Brothers Limited completed an audit of the company's books. In addition, there was a second meeting between Mr. Jones and Ms. Finnigan. At the meeting, Mr. Jones strongly recommended that, if Acme Ltd. did, in fact, purchase Jones Brothers Limited, Acme ought to seriously consider entering into a lease of desirable warehouse space located next door to Jones Brothers in order to accommodate Jones Brothers' expanding inventories. Mr. Jones emphasized that he had discussed the lease with the landlord; the landlord had informed him that, if a lease were to be signed, it must be signed not later than January 29, 1995. Eventually, the president of Jones Brothers Limited wrote the following letter to Acme Ltd.:

Acme Ltd.,   January 25, 1995
Toronto, Ontario.

Attention: Jane Finnigan

Dear Ms. Finnigan:

Jones Brothers Ltd. hereby agrees to sell to Acme Ltd. all of the assets, undertaking and business of Jones Brothers Limited for a total purchase price of $1,000,000, subject to the terms and conditions of the attached form of agreement, to which is attached a current audited financial statement of Jones Brothers Limited.

This offer shall be firm and irrevocable for a period of twenty days from the date of this letter and your agreement to purchase our business in accordance with these terms and conditions will be sufficiently evidenced by your signing the enclosed copy of this letter. Upon receipt of your acknowledgement we can then arrange a meeting for the purpose of executing the attached agreement.

Yours very truly,
Jones Brothers Limited

Mr. Jones personally delivered the letter to Ms. Finnigan on January 25, 1995.

On January 28, 1995, Mr. Jones mailed a letter to Ms. Finnigan revoking the offer to sell the business to Acme.

On January 29, 1995 Ms. Finnigan, as president of Acme Ltd., executed a two-year lease of warehouse space next door to Jones Brothers Limited. The warehouse was the building that Mr. Jones and Ms. Finnigan had previously discussed. Had it not been for Acme's interest in purchasing Jones Brothers Limited, Ms. Finnigan would not have executed the lease.

Ms. Finnigan received Jones's revocation on January 31, 1995. Immediately after receiving the revocation, Ms. Finnigan signed the duplicate of the letter dated January 25, 1995, and personally delivered it to Mr. Jones. Ms. Finnigan adamantly insisted that, in the circumstances of all of the correspondence and the meetings that had taken place between them, Jones Brothers Limited was obligated to either sell its business to Acme Ltd., or to assume the lease obligations and make all payments to the landlord.

a. Explain the nature of the legal relationships between Acme Ltd. and Jones Brothers Limited when:

> i. the letter of December 28, 1994 had been signed and returned by Mr. Jones;
> ii. the letter of January 25, 1995 had been signed and returned by Ms. Finnigan.

b. Discuss the merits of Ms. Finnigan's claim that Jones Brothers Limited was obligated to sell its business or assume the warehouse-lease obligations.

Commentary:

a.(i) In answering, comment on whether the letter of December 28, 1994 constituted an enforceable contract? If not, what was it?

 (ii) In commenting on the effect of the letter of January 25, 1995, point out the "mailbox" doctrine relating to acceptance and revocation and apply the rules to the facts.

b. Your answer in (ii) will be determinative here. As much as you might want to sympathize, remember the significance of the contract rules. Point out what Ms. Finnigan ought to have done.

CHAPTER NINE

INTENT

MUTUAL INTENT

The engineer should make sure that any contract document specifies the agreement between the parties on all essential terms.

LETTERS OF INTENT

The use of "letters of intent" in business is a very common practice. Businesses use the letter to express interest in proceeding with a particular transaction, usually on the basis of further negotiation and subsequent agreement. Sometimes letters of intent are clearly agreements to agree, rather than well-defined agreements. The agreements to agree do not constitute enforceable contracts: the courts will not enforce an agreement to agree. It is, in fact, no agreement at all.

To illustrate: in 1976, the Ontario Court of Appeal heard the case of *Bahamaconsult Ltd.* v. *Kellogg Salada Canada Ltd.*[1] There were certain omissions in a letter of intent relating to the sale of the shares of a company. The court noted:

> The trial Judge found that the document of October 10, 1969, was a contract complete in itself, and that while it was the intention of the parties to draw a further agreement, the subsequent agreement was only to spell out the mechanics of the transfer of the shares, and of the closing of the sale. With respect, we think that the trial Judge erred in so finding. We are all of the opinion that the document of October 10th does not contain certain essential terms, and that it was the intention of the parties that these terms would be negotiated between them and embodied in a subsequent agreement. Since the parties

[1] 75 D.L.R. (3d) 522

were unable to agree upon those terms, there was no enforceable contract. The applicable principle of law is stated in the oft-quoted words of Parker, J., in *Von Hatzfeldt-Wildenburg* v. *Alexander*, [1972] Ch. 284 at pp. 288-9:

> It appears to be well settled by the authorities that if the documents or letters relied on as constituting a contract contemplate the execution of a further contract between the parties, it is a question of construction whether the execution of the further contract is a condition or term of the bargain or whether it is a mere expression of the desire of the parties as to the manner in which the transaction already agreed to will in fact go through. In the former case there is no enforceable contract either because the condition is unfulfilled or because the law does not recognize a contract to enter into a contract. In the latter case there is a binding contract and the reference to the more formal document may be ignored.

The court noted that both parties had referred to the document in question as "a letter of intent." Essential terms of the contract were missing; those terms could only be agreed on by further negotiation between the parties.

A letter of intent that does not contain all essential terms is not an enforceable contract. It can, however, serve a useful purpose: it may establish terms for negotiation and it may create some moral obligation between the parties to continue to negotiate in good faith.

To the inexperienced, the letter of intent can present considerable difficulties. A letter of intent could constitute an agreement, if it were sufficiently detailed. The engineer should be cautious when presented with "letters of intent"; he or she should consult a lawyer if in doubt as to the true nature of the letter. The precise drafting of contracts is better left to lawyers. The engineer in business can greatly simplify the contract preparation process by listing negotiated business terms, but it is advisable to seek legal advice when negotiating significant terms and when drafting the contract document.

CHAPTER TEN

CONSIDERATION

CONSIDERATION

Consideration is an essential part of an enforceable contract. As *Black's Law Dictionary* points out, it is the cause, motive, price, or impelling influence that induces a contracting party to enter into a contract. Consideration can be described as something of value that is exchanged by contracting parties. A promise made by an engineer to design a structure in return for the payment of a fee is an example of a situation where consideration exists. Each party to the contract promises something in return for the other party's undertakings. The payment of money is not essential: consideration may consist of an exchange of promises, each promise representing something of value.

The courts are not normally concerned with the adequacy of consideration. There are exceptional circumstances, however. For example, if it is established that the contract was entered into under conditions amounting to undue influence, duress, or fraud, the courts will provide relief.

Where consideration is not present in the form of promises or other mutual exchange of something of value, no contract is formed unless the document is "sealed." There are two kinds of seals. A mechanical device is used to imprint corporate seals on documents executed by corporations; the personal seal of the individual is a small red adhesive wafer. The ancient practice remains with us today and is recognized by the courts as a substitute for consideration. Originally, the use of the seal was considered an act of great importance, a clear indication that the promisor intended to be bound by his promise. The party making the promise is required to affix the seal to the document expressing the promise or undertaking.

The use of the seal is important in tendering procedures. Often tenders submitted by bidding contractors are required to be irrevocable for a specific period of time; for example, an offer might be binding for twenty days following the date of the opening of the tenders. An "irrevocable" offer without consideration or a seal is simply a gratuitous promise; it is not legally binding. The offeror may revoke the offer at any time before its acceptance. When the offeror promises to hold an offer open for a specified period, separate consideration is required for the promise of irrevocability to be binding. This separate consideration is usually achieved through the use of a seal.

EQUITABLE ESTOPPEL

But should a party that makes a gratuitous promise (that is, a promise without consideration) be entitled to escape its moral obligations? Such obligations are strictly moral, and are not legally binding, but can still raise questions of an equitable nature. An example is the 1963 case, *Conwest Exploration Co. Ltd. et al.* v. *Letain,*[1] a decision of the Supreme Court of Canada. An option agreement that related to certain mining claims owned by the optionor had a time limit. The optionee had to take certain steps by a specified date in order to be entitled to exercise the option to acquire the mining claims. Before the option's expiry, the optionor became aware that the optionee would not be able to fulfil his obligations by the expiry date. The optionor implied that the time for fulfilment was extended. However, the promise to extend was not accompanied by consideration; hence, it was not strictly binding. Subsequently, the optionor reverted to his strict contractual rights and insisted that the original expiry date must apply. The Supreme Court of Canada held that it would be inequitable if the optionor were permitted to enforce the original contract in the circumstances and that the optionor should therefore be "estopped" from reverting to his strict contractual rights.

The *Conwest* case is very important. Remember that, pursuant to contract law, consideration (or a seal) must be present in order to make an amendment to a contract enforceable — otherwise the amending promise is gratuitous. *Conwest* provides a basis on which to argue that where the terms of a contract are amended without the consideration that would make the amending promise enforceable, there may be relief for the party that relies upon the gratuitous promise. The concept whereby such relief may be pro-

[1]   41 D.L.R. (2d) 198

vided is called "promissory" or "equitable estoppel." A court will only exercise its discretion to apply the concept of promissory or equitable estoppel, however, to avoid an obviously inequitable result.

The concept of equitable estoppel has been examined in other cases. The 1968 decision by the Supreme Court of Canada in *John Burrows Ltd.* v. *Subsurface Surveys Ltd. et al.*[2] is an example. The plaintiff sought to enforce the terms of a promissory note. The terms were as follows:

> FOR VALUE RECEIVED Subsurface Surveys Ltd. promises to pay to John Burrows Ltd. or order at the Royal Bank of Canada the sum of Forty-two Thousand Dollars ($42,000.00) in nine (9) years and ten (10) months from April 1st, 1963, payable monthly on the first day of May, 1963, and on the first day of each and every month thereafter until payment, provided that the maker may pay on account of principal from time to time the whole or any portion thereof upon giving thirty (30) days' notice of intention prior to such payment.
>
> In default of payment of any interest payment or instalment for a period of ten (10) days after the same became due the whole amount payable under this note is to become immediately due.

> SUBSURFACE SURVEYS LTD.
> [Sgd.] G. Murdoch Whitcomb
> President

Although payments were made late, the holder of the note did not insist on all the terms. Continuing indulgences were granted to Subsurface Surveys Ltd. Eleven payments were accepted more than ten days after they were due. The parties eventually had a falling out. When one of the interest payments was thirty-six days overdue, the holder of the note decided to insist on the default clause. Subsurface Surveys was notified that immediate payment of the principal amount must be made, in accordance with the terms of the note.

Subsurface Surveys protested on the basis that the noteholder should be equitably estopped from enforcing its strict contractual rights in all of the circumstances of case. Subsurface Surveys argued that the holder of the note was contradicting an implied

[2] 68 D.L.R. (2d) 354

agreement between the two parties. The default clause of the note had been disregarded several times by the noteholder; Subsurface Surveys inferred that they had an agreement with the noteholder with respect to the default clause. The Supreme Court of Canada concluded that this evidence did not warrant such an inference.

In the *John Burrows* decision, the Supreme Court of Canada referred to the following statements of earlier court decisions on the principles involved:

> ... it is the first principle upon which all Courts of Equity proceed, that if parties who have entered into definite and distinct terms involving certain legal results — certain penalties or legal forfeiture — afterwards by their own act or with their own consent enter upon a course of negotiation which has the effect of leading one of the parties to suppose that the strict rights arising under the contract will not be enforced, or will be kept in suspense, or held in abeyance, the person who otherwise might have enforced those rights will not be allowed to enforce them where it would be inequitable having regard to the dealings which have thus taken place between the parties ...
>
> The principle, as I understand it, is that where one party has, by his words or conduct, made to the other a promise or assurance which was intended to affect the legal relations between them and to be acted on accordingly, then, once the other party has taken him at his word and acted on it, the one who gave the promise or assurance cannot afterwards be allowed to revert to the previous legal relations as if no such promise or assurance had been made by him, but he must accept their legal relations subject to the qualification which he himself has so introduced, even though it is not supported in point of law by any consideration, but only by his word.

Justice Ritchie of the Supreme Court then stated in the *John Burrows* case:

> It seems clear to me that this type of equitable defence can not be invoked unless there is some evidence that one of the parties entered into a course of negotiation which had the effect of leading the other to suppose that the strict rights under the contract would not be enforced, and I think that this implies that there must be some evidence from which it can be inferred

that the first party intended that the legal relations created by the contract would be altered as a result of the negotiations.

It is not enough to show that one party has taken advantage of indulgences granted to him by the other for if this were so in relation to commercial transactions such as promissory notes it would mean that the holders of such notes would be required to insist on the very letter being enforced in all cases for fear that any indulgences granted and acted upon could be translated into a waiver of their rights to enforce the contract according to its terms.

On the facts, the *John Burrows* case was distinguished from the *Conwest* case.

In 1979, the Ontario Court of Appeal decided *Owen Sound Public Library Board* v. *Mial Developments Ltd. et al.*,[3] in which the issue of equitable estoppel was raised in an action for breach of contract. The contract in question was a construction contract, which provided that payments were to be made by the owner within five days of an architect's certificate. If the owner should fail to pay any sum certified as due by the architect within seven days, the contractor would be entitled to terminate the construction contract with the owner. The architect had certified such a sum as due. Soon after that, the parties agreed upon certain action; as a result, the owner assumed that the due date for payment to the contractor was being extended. Instead of making a payment, the owner had requested the contractor to have the corporate seal of one of its subcontractors affixed to a document supporting the architect's certificate. The contractor undertook to obtain the corporate seal. The payment date passed, but the contractor did not obtain the corporate seal. The contractor later purported to terminate the contract because of the owner's failure to pay within the time limit. The Court of Appeal concluded that the contractor's conduct had led the owner to believe that the time limit for payment would be extended until the subcontractor's sealed document had been provided. The court concluded that the owner's assumption was reasonable and held that the contractor should be estopped from invoking his strict contractual termination rights. Enforcement of contractual rights would be clearly inequitable. The contractor was attempting to take advantage of the owner's contractual default, but that default had been induced by the contractor's conduct.

[3] 26 O.R. (2d) 459

The issue of equitable estoppel can be of particular significance to the engineer who is acting as a contract administrator. If an engineer waives any particular contractual rights, the engineer may be faced with the argument that the engineer (or the party on whose behalf the engineer is acting) ought subsequently to be equitably estopped from reverting to his or her strict contractual rights. The success of the argument will depend upon the particular circumstances of each case.

SAMPLE CASE STUDY

The following hypothetical case and sample answer is included for illustrative study purposes.

A mining contractor signed an option contract with a landowner, which provided that if the mining contractor (the "optionee") performed a specified minimum amount of exploration services on the property of the owner (the "optionor") within a nine-month period, then the optionee would be entitled to exercise its option to acquire certain mining claims from the optionor.

Before the expiry of this nine-month "option period," the optionee realized that it couldn't fulfil its obligation to expend the required minimum amount by the expiry date. The optionee notified the optionor of its problem prior to expiry of the option period and the optionor indicated that the option period would be extended. However, no written record of this extension was made, nor did the optionor receive anything from the optionee in return for the extension.

The optionee then proceeded to perform the services and to finally expend the specified minimum amount during the extension period. However, when the optionee attempted to exercise its option to acquire the mining claims the optionor took the position that, on the basis of the strict wording of the signed contract, the optionee had not met its contractual obligations. The optionor refused to grant the mining claims to the optionee.

Was the optionor entitled to deny the optionee's exercise of the option? Explain the contract law principles that apply to the positions taken by the optionor and by the optionee.

Sample Answer:

This is a contract law case involving the principle of equitable estoppel. Equitable estoppel is a concept that can be applied to prevent a party to a contract from enforcing its strict contractual terms in circumstances where doing so would produce an unfair and inequitable result. It can be applied where the parties to a contract have negotiated other terms, even though they might not have formally amended the original contract as clearly as they should have.

Here the optionor indicated that the option period would be extended but did so without receiving any consideration for the extension. Arguably, therefore, the optionor's consent was but a gratuitous promise. Gratuitous promises, lacking consideration, do not constitute legally binding contracts, or in this case, a legally binding amendment agreement.

Normally, rights set out in a contract are enforceable in accordance with the terms of the contract. Here, however, given what occurred between the parties — which was in effect a negotiation that led the optionee to reasonably assume that the optionor was not going to enforce its strict rights under the contract — the court would stop the optionor from enforcing the rights under the strict wording of the contract. To enforce the contract would result in an inequitable and unfair outcome.

These case facts strongly resemble an actual case on this issue that came before the Canadian courts. (Note: reference to the case by name, *Conwest*, is optional.)

(Further note: although some examination candidates refer to the parol evidence rule in answering this case, it is not an appropriate reference insofar as the parol evidence rule deals with terms of a contract that are discussed before the written contract but omitted when the written contract is subsequently prepared. Here, the negotiated, but gratuitous, extension occurred after the written contract.)

CHAPTER ELEVEN

CAPACITY

MINORS

In order for a contract to be binding and enforceable, all parties must have the necessary capacity to enter into a contract. Under the common law, not everyone has the "necessary capacity." For example, a contract with a minor is enforceable by the minor but unenforceable by the other party, unless it can be established that the contract concerned something that was necessary to the minor (for example, food, clothing, shelter, and so on) or unless the contract is ratified by the minor upon reaching the age of majority. The age of majority is now eighteen in Ontario, Manitoba, Saskatchewan, Alberta, and Prince Edward Island; it is nineteen in Nova Scotia, New Brunswick, and Newfoundland.

The age of majority is also nineteen in British Columbia. However, the Infants Act[1] of that province prevents the ratification of a contract made by a minor. That is, a contract made by a minor will generally remain unenforceable even if the minor attempts to ratify the contract after he or she has reached the age of majority.

DRUNKS AND LUNATICS

Contracts for non-necessaries entered into by lunatics or intoxicated persons are enforceable by the lunatic or drunkard but unenforceable by the other party — on two conditions. The other party to the contract had to be aware of the state of insobriety or lunacy. And the incapacitated party must repudiate the contract within a reasonable period of time. Anyone who claims to have signed a contract while inebriated will obviously have some very significant evidentiary difficulties: it must be substantiated that the individual was drunk when entering into the contract, and it

[1] R.S.B.C. 1979, c. 196

must be shown that the other party appreciated that the individual was drunk.

CORPORATIONS

Caution should be exercised in dealing with corporations. The engineer must make sure that it is within the powers of the contracting corporation to carry out the obligations described in the contract. If it is clearly beyond the power of the corporation to enter into the contract, the contract will not be enforceable. It is particularly important, when dealing with corporations that have been created by special statute of the federal or provincial legislature (such as a railway company, for example), to determine the nature of the corporation. Its incorporating statute must provide that the purpose of the proposed contract is within the powers of the corporation. During negotiations, it is advisable to require that the corporation expressly represent, in the contract, that it has the necessary capacity to enter into the contract, and that the contract will be enforceable against it.

In day-to-day business dealings with a corporation, reliance upon representatives of the corporation is appropriate as established in the case of *Royal British Bank* v. *Turquand*, an English case reported in 1856. A corporation is bound by the acts of its officials, provided such acts are within the actual, usual, or apparent scope of each such official's authority, and the party dealing with the corporation has no knowledge to the contrary, and provided there are no suspicious circumstances or prohibitions in the corporation's public documents. If there is any doubt about the authority of an official to act on behalf of a corporation, appropriate enquiries should be made, including a review of the corporation's public file. The file will include the corporation's charter documents and a listing of its directors and officers. A file is kept by the appropriate government department for each corporation.

CHAPTER TWELVE

LEGALITY

CONTRARY TO STATUTE LAW

A contract will not be enforced if the purpose of the contract is unlawful, that is, if it is illegal or void because it is contrary to any statute.

There are many examples of contracts contrary to statutory law. They include:

1. a contract contrary to the provisions of the Bankruptcy Act (Canada).[1] For example, the Act provides that where property is transferred between related parties and the transferor becomes bankrupt within one year of the transfer, the transfer is void as far as the Trustee in Bankruptcy is concerned. The Act also provides that if, within three months before bankruptcy, an insolvent party transfers property to a creditor with the intent of giving that creditor a preference over other creditors, the transfer is fraudulent and void;

2. a contract that is contrary to provincial workers' compensation legislation. For example, section 19 of the Workers' Compensation Act of Ontario[2] provides that a contract between an employer and worker is invalid if it fixes an amount of compensation in lieu of compensation pursuant to the Act;

3. a contract that is contrary to the provisions of the Competition Act (Canada).[3] For example, a contract might represent an attempt to prevent or unduly lessen competition, or to engage in "bid-rigging";

---

[1] R.S.C. 1985, c. B-3
[2] R.S.O. 1990, c. W-11
[3] R.S.C. 1985, c. C-34

4. a contract that provides for a waiver of lien rights contrary to section 4 of the Construction Lien Act of Ontario; R.S.O. 1990, c. C. 30;

5. a contract for services where the party to perform is required to be licensed pursuant to a statute or by-law. Failure to license may expressly preclude the right to contract for such services. However, as will be noted, case decisions to date have varied.

An illustration of Example 5 is the 1958 decision of *Kocotis* v. *D'Angelo*.[4] The Ontario Court of Appeal heard the case. An electrician began an action for payment for work done and material supplied. The electrician was not properly licensed as an electrical contractor pursuant to a local by-law. In its decision, the court stated:

> It is plain to me that the object of the by-law was to protect the public against mistakes and loss that might arise from work done by unqualified electricians. It was not to secure the revenue from certificates or from licenses, because only certain qualified persons could obtain such certificates or licenses. It was plainly intended by the by-law to prohibit a maintenance electrician from undertaking the work of a master electrician or electrical contractor, and no maintenance electrician could lawfully contract for any electrical work.... Such a contract would be illegal and could not be enforced in the Courts.

The *Kocotis* decision was followed in the 1974 case of *Calax Construction Inc.* v. *Lepofsky*.[5] The case involved an unlicensed building contractor. The Ontario High Court of Justice concluded that the contract was illegal and unenforceable. The court referred to an excerpt from Cheshire and Fifoot, *Law of Contract*, at page 334, 8th edition (1972):

> The general principle, founded on public policy, is that any transaction that is tainted by illegality in which both parties are equally involved is beyond the pale of the law. No person can claim any right or remedy whatsoever under an illegal transaction in which he has participated.

[4]   [1958] O.R. 104
[5]   [1974], 5 O.R. (2d) 259

But the 1979 decision in *Monticchio* v. *Torcema Construction Ltd. et al.*[6] represented a departure from the precedents. The Ontario High Court of Justice reviewed the earlier decisions on the question of licensing. The court pointed out that because the plaintiff was not a licensed drain contractor, the contract itself might be illegal. But the court felt that that was not a complete defence against the contractor's claim for payment. The by-law required only that the contractor be licensed; there was no prohibition against the sale of material, thus the court held that the contractor ought to at least be paid for material supplied. The judgment also indicated that a claim for payment for services on a time basis might succeed even though the contract was held to be void. The *Monticchio* case suggests a change in the attitude of the court with respect to compensation for unlicensed contractors. Whether such change will be endorsed generally by the courts remains to be seen.

CONTRARY TO COMMON LAW

A contract that contravenes statutory law may be illegal and/or void; a contract that is against public policy may be illegal and/or void according to common law.

Contracts may contain provisions that are against public policy. An example is a contract that contains restrictive covenants in restraint of trade. The court presumes initially that any agreement in restraint of trade is against public policy, and therefore void. However, a party seeking to enforce a restrictive covenant can overcome the court's initial presumption. The party must prove that, between the parties to the contract, the restrictive covenant is a reasonable one, and that it does not adversely affect the public interest.

An example is a contract for the purchase and sale of a business. The purchaser may require the vendor to covenant not to compete with the purchaser in a similar business in a specified area and over a specified period of time. Such a provision is generally referred to as a "non-competition agreement." Such an agreement might read:

> The vendor shall not for a period of five years from the date of closing, anywhere within the Province of Ontario, either alone or in conjunction with any individual, firm, corporation,

[6] 26 O.R. (2d) 305

association or other entity, whether as principal, agent, shareholder, employee or in any other capacity whatsoever carry on, or be engaged in, concerned with or interested in, directly or indirectly, any undertaking similar to any of the business carried on by the company being purchased within the respective territories in which such business is then carried on.

A purchaser seeking to enforce a non-competition agreement must persuade the court that the terms are reasonable between the parties, and that the nature of the vendor's services are such that the public interest would not be adversely affected if the covenant were enforced. There is no way of predicting what the court will determine to be "reasonable" in terms of time and geographic area. Each case will turn on its own facts. If the court is not persuaded that the terms are reasonable, the restrictive clause will not be enforced. If the court determines that the provisions are unreasonable, it will not intervene to enforce more reasonable terms.

Engineers often encounter similar restrictive covenants in employment contracts. Such contracts may attempt to restrict the employee after leaving such employment. The courts will apply principles of reasonableness and public policy to determine the enforceability of non-competition clauses in employment contracts. The courts are reluctant to enforce restrictive covenants that would severely limit the former employee's ability to earn a livelihood; each decision will depend upon the particular circumstances.

SAMPLE CASE STUDY

The following hypothetical case and commentary is included for illustrative study purposes.

1. A professional engineer entered into a written employment contract with a Toronto-based civil-engineering design firm. The engineer's contract of employment stated that, for a period of five years after the termination of the employment, the engineer would not practise professional engineering either alone, or in conjunction with, or as an employee, agent, principal, or shareholder of an engineering firm anywhere within the City of Toronto.

During the engineer's employment with the design firm, the employee engineer dealt directly with many of the firm's clients. The engineer became extremely skilled in preparing cost estimates, and established a good reputation within the City of Toronto.

The engineer terminated employment with the consulting firm after three years, and immediately set up an engineering firm in another part of the City of Toronto. The engineer's previous employers then commenced a court action for an injunction, claiming that the engineer had breached the contract and *should not* be permitted to practise within the City limits.

Do you think the engineer's former employers should succeed in an action against the engineer? In answering, state the principles a court would apply in arriving at a decision.

Commentary: 1. In answering, point out that an agreement in restraint of trade, such as this, is only enforceable if it meets the tests of reasonableness with respect to both time and geographic area and public policy tests described in the chapter. Explain what those tests are and then analyze the facts and apply the tests to support your conclusion.

CHAPTER THIRTEEN

THE STATUTE OF FRAUDS

A contract may be verbal or written. Written contracts may be formed through correspondence. A contract can also be formed in part by correspondence and in part by discussions between the parties.

An agreement between parties is best set out in the form of a single written contract that clearly details the agreement. Such a contract may, in fact, be essential in order to ensure that it is enforceable; the statute of frauds (of the various common-law provinces) stipulates that certain types of contracts must be in writing to be enforceable.

For example, section 4 of the Statute of Frauds of Ontario[1] states:

> 4. No action shall be brought whereby to charge any executor or administrator upon any special promise to answer damages out of the executor's or administrator's own estate, or to charge any person upon any special promise to answer for the debt, default or miscarriage of any other person, or to charge any person upon any contract or sale of lands, tenements or hereditaments, or any interest in or concerning them, or upon any agreement that is not to be performed within the space of one year from the making thereof, *unless the agreement upon which the action is brought, or some memorandum or note thereof is in writing and signed by the party to be charged* therewith or some person thereunto lawfully authorized by the party.

Various types of contracts are referred to in the Statute of Frauds. Those most likely to be relevant to the engineer are:

1. contracts relating to interests in land (that is, "any contract or sale of lands, tenements or hereditaments, or any interest in or concerning them");

[1] R.S.O. 1990, c. S. 19

2. those agreements that are not to be performed within the space of one year from the making thereof; and

3. guarantees of indebtedness.

Agreements relating to ownership interests and leasehold interests in land must be in writing to be enforceable. Contracts for the construction of buildings, on the other hand, need only be in writing if the contract cannot be performed within a year, although a written contract is certainly preferable to an oral one. The engineer should ensure that legal advice is obtained regarding interests in real property — land — affected by agreements. Real-property law is a matter that is most appropriately handled by lawyers.

A contract between an engineer and client is not usually a contract that must be in writing pursuant to the Statute of Frauds. Complications may arise, however, if performance by either party cannot be completed within one year.

A contract of guarantee must also be in writing in order to be enforceable. A distinction between a guarantee and an indemnification was discussed earlier in the text.[2] There is a further distinction between the two: an indemnification need not be in writing to be enforceable. In some circumstances, however, it may be difficult to distinguish between a contract of guarantee and one of indemnification. It is advisable to put both kinds of agreement in writing.

DERIVATION OF STATUTE

A statute of frauds is in force in each of the various common-law provinces of Canada. The statute is derived from the English Statute of Frauds. One purpose of the original statute was to answer concern about property interests. The statute was developed to prevent property interests being lost through fraudulent testimony about verbal agreements to convey property interests.

UNENFORCEABLE CONTRACTS NOT VOID

Although a verbal contract may be unenforceable because of the Statute of Frauds, it will not be treated as void: the courts will recognize its existence for certain purposes. For example, sup-

[2] Supra, at page 9.

pose that someone does not honour a verbal agreement to convey real property. The courts will not permit the defaulting party to retain a deposit cheque. The verbal agreement will be recognized so that the deposit may be recovered by the non-defaulting party.

DESIRABILITY OF WRITTEN FORM

There is an obvious problem with any verbal agreement: it may be extremely difficult to substantiate the terms of the contract. This difficulty is best avoided by ensuring that all contracts are in written form.

CHAPTER FOURTEEN

MISREPRESENTATION, DURESS, AND UNDUE INFLUENCE

MISREPRESENTATION

A misrepresentation is a false statement or assertion of fact. If a misrepresentation is made to induce a party to enter into a contract, the misled party may apply to the court to have the contract rescinded. The court will treat the contract as voidable at the option of the party misled. When a contract is rescinded, it is cancelled or set aside.

An innocent misrepresentation is a false assertion made by a party who does not appreciate that the statement is false.

A fraudulent misrepresentation has been described, by the English Court of Appeal in *Derry* v. *Peek*,[1] as a statement made "(1) knowingly, or (2) without belief in its truth, or (3) recklessly, careless whether it be true or false." The court noted that a party who makes a careless statement can have no real belief in the truth of what he or she states.

If a person deceived by a misrepresentation has entered into a contract, there are remedies available. The choice of remedy depends upon the kind of misrepresentation: innocent or fraudulent.

An innocent misrepresentation is remedied by rescission of the contract. The deceived party must repudiate the contract within a reasonable length of time. The deceived party is also generally entitled to claim compensation for damages in respect of any costs sustained as a result of entering into the contract.

Where the misrepresentation is fraudulent, a deceived party is generally entitled to rescind the contract and to claim compensation for reasonable costs incurred as a result of entering into the contract. The party can also sue for damages for deceit.

[1] [1889] 14 CAS. 337

In addition, the *Hedley Byrne* decision has established potential liability for a negligent misrepresentation or negligent misstatement.

MISREPRESENTATIONS IN ENGINEERING SPECIFICATIONS

Contractors rely on engineering plans and specifications. If those plans contain misrepresentations, a contractor may be entitled to rescind the construction contract. An example is *Township of McKillop* v. *Pidgeon and Foley*.[2] The defendant contractor had submitted its tender based on a price estimated by an engineer. Later, the contractor discovered that considerably more excavation had to be done than estimated. (The job required approximately 16 percent more work than had been estimated.) Because of the error in the specifications, the contractor terminated its contract; the contractor then sued for breach of contract. The court heard evidence from a number of witnesses, including civil engineers and experienced surveyors. The court concluded that the error of 16 percent was excessive and would practically deprive the contractor of any profit; 16 percent might result in a loss to the contractor, in circumstances where he should not be asked to incur such loss. As the trial judge noted, contracts frequently stipulate that architects' or engineers' estimates are not binding upon the property owner. The judge noted that errors in estimates should not entitle the contractor to any further money and that, when contractors had sued in such cases, they had occasionally failed. However, the judge stated that there was no doubt the contractor had entered into the contract upon the faith of the estimate. It would be unreasonable, the judge stated, to expect the contractor to do the work of the engineer. The judge concluded that no deceit was involved; instead, a mistake had arisen from an innocent misrepresentation in the tendering documents. The defendant contractor was thus entitled to repudiate the contract.

DURESS

If a contract is induced by means of intimidation, it is voidable. Such intimidation is termed "duress." Duress can be defined as

[2]   [1908] 11 O.W.R. 401

threatened or actual violence or imprisonment used as a means of persuading a party to enter into a contract. The actual or threatened violence or imprisonment must be directed at the contracting party or a close relative. For example, lawful imprisonment — as the result of criminal prosecution — might be threatened. The court may nevertheless provide equitable relief. The case of *Mutual Finance Co. Ltd.* v. *John Wetton & Sons Ltd.*[3] is an example. A family member had forged a previous guarantee. Disclosure was threatened in an attempt to execute a second guarantee. At the time of the coercion, the party threatening disclosure knew that the alleged forger's father was in ill health, and that the shock of the disclosure might kill him. The guarantee was held to be unenforceable.

ECONOMIC DURESS

An interesting 1993 decision of the British Columbia Supreme Court dealt with the subject of "economic duress." The case was *Gotaverken Energy Systems Ltd.* v. *Cariboo Pulp & Paper Co.*,[4] a case involving a contract to retrofit a recovery boiler in a pulp mill. The contractor agreed for a price in excess of $26,000,000 to retrofit the recovery boiler, on a "turnkey" project basis during one of the mill's shut-down periods. The contractor was to work two eleven-hour shifts, seven days a week. After the contractor had been paid more than $24,000,000, problems had developed that led to work interruptions that threatened the contractor with substantial losses. A major part of the problem were occurrences of "gas outs" that affected the contractor's personnel, some of whom had to be hospitalized. The owner admitted responsibility for the work stoppages caused by the "gas outs" and undertook to compensate the contractor. However, the contractor threatened to reduce its work schedule to $37\frac{1}{2}$ hours per week in the circumstances unless the owner agreed to amend the contract from a fixed price contract to a time and material basis contract. During the negotiating meetings, the owner's representative often referred to the fact that the contractor was "holding a gun to the head of the owner," as the threatened change in work schedule would have caused severe economic loss to the owner. The court concluded that economic duress was involved in persuading the owner to enter into the time and materials amending agreement,

[3] [1937] 2 K.B. 389
[4] 9 C.L.R. (2d) 71

to its economic detriment. The court held that such pressure on the owner was not legitimate and therefore the amending agreement was not valid as it had been entered into under economic duress. Accordingly, the contractor did not recover the full amount it had claimed in the lawsuit, but it did recover a substantial amount on account of extras under the original contractual arrangement. The contractor had sued for an amount in excess of $10,000,000 and was ultimately awarded damages totalling over $6,000,000.

In considering the law on economic duress, the court made reference to U.K. precedents, in particular a 1989 decision of the English courts in *Atlas Express Ltd.* v. *Kafco (Importers & Distributors) Ltd.*,[5] with respect to which the court pointed out:

> That was a situation more akin to this case, because it is an example of a party to a contract being forced to renegotiate the terms to his disadvantage. There was no alternative but to accept the new terms offered. In holding there had been economic duress which vitiated the apparent consent to new terms, the case stands as clear authority for the proposition that a new agreement will not be enforced if its terms are accepted under duress and where it is revoked after the illegitimate pressure has ceased to operate.

The court also emphasized that in the particular circumstances of this contract, there was no evidence any other firm was even available to do this work. As the court pointed out:

> In my view, it was not practical to even contemplate the idea of shutting down the retrofit work and seeking another contractor to finish the job. There was no pool of contractors willing and able to take on a job of this type. Even if one could be found, one can only speculate on how long such an exercise would take. As well, it would mean a time and material contract, a situation with no saving, involving considerable loss occasioned by down time which Bender understood could not be recovered.

The *Gotaverken Energy Systems* case provides an interesting illustration of the application of the concept of duress in a commercial contracting situation. Exceptional circumstances are required to sustain the argument of economic duress, particularly

[5]   [1989] 1 All E.R. 641

the fact that at the time of being coerced into making the contract no alternative course of action was available, and the party succumbing to the coercion had no alternative but to accept the proposed contractual arrangement or suffer severe economic loss. Where that can be substantiated, the *Gotaverken Energy* case supports the principle that such a contact procured under economic duress is not valid.

UNDUE INFLUENCE

Undue influence is similar to duress, but arises in less drastic circumstances. Undue influence occurs where one party to a contract dominates the free will of the other party to such an extent as to be able to coerce the dominated party into an unfair agreement. In such circumstances the dominated party is entitled to be relieved of contractual obligations. Undue influence is an equitable concept. It is not frequent in business situations, where parties are at arms' length. It allows family members (for example, husband and wife or parent and minor child) to repudiate a contract where bargaining positions are unequal and undue influence occurs.

CHAPTER FIFTEEN

MISTAKE

The subject of mistake in contract is of interest to engineers. It is often raised in connection with the submission of bids during the construction tendering process.

The common law has long recognized that it may be equitable to provide some relief if a mistake is made by one or both of the parties to a contract. But the courts will intervene to provide relief to a contracting party that has made a mistake only in rare circumstances.

RECTIFICATION

If contracting parties have clearly reached agreement but have recorded the provisions of the agreement inaccurately in a written contract, a "common mistake" has occurred. One of the parties to the agreement can apply to the court for an order of rectification. The order is used to correct an obvious common mistake. The party applying for the order must persuade the court that the written contract is inconsistent with the terms agreed upon by the parties; the mistake must be of a secretarial or recording nature.

UNILATERAL MISTAKE

A unilateral mistake is a mistake made by only one party to a contract. Unilateral mistakes by contractors in tendering have resulted in several interesting case decisions.

For example, in 1960 the British Columbia Court of Appeal heard the case of *Imperial Glass Ltd.* v. *Consolidated Supplies Ltd.*[1] The contractor (or offeror) had used the wrong figure in

[1]   22 D.L.R. (2d) 759

calculating the price at which it would supply certain items. The offeree was aware that the offeror had made the mistake; however the offeror had not been induced to make the mistake by any representation of the offeree. The court was satisfied that the offeree's conduct was not fraudulent. The offeree's conduct might be open to question on moral or ethical grounds, but the court would not relieve the contractor from the consequences of his mistake.

But subsequent decisions of Ontario courts held that a contractor may be relieved of the consequence of unilateral mistake in certain circumstances.

An example is the 1977 decision of the High Court of Justice of Ontario in *Belle River Community Arena Inc.* v. *W.J.C. Kaufmann Co. et al.*[2] In submitting a bid, the defendant contractor had incorrectly transferred a figure from a summary sheet. The contractor's bid was therefore approximately $70,000 lower than intended. The total bid price was $641,603. The bid was submitted under seal, and was irrevocable for sixty days. When the contractor discovered his mistake, he attempted to withdraw his tender. There was no disagreement that the error had been made. The plaintiff refused to allow the contractor to withdraw the tender. More than a month after being informed of the mistake, the plaintiff attempted to accept the contractor's tender. The Ontario court was critical of the plaintiff's motive for accepting the tender. In finding in the contractor's favour, the court pointed out that the plaintiff had not submitted to the contractor a formal contract executed by the plaintiff. Hence the plaintiff had not technically obtained an unequivocal refusal from the contractor to enter into the contract. The plaintiff entered into a contract with another tenderer, then sued the defendant contractor for the difference between the amounts of the two tenders. Before deciding the case, the court considered the British Columbia Court of Appeal decision in *Imperial Glass Ltd.* v. *Consolidated Supplies Ltd.*[3] The court noted a distinction in the nature of the mistake made in the two cases. In the *Imperial Glass* case, the court pointed out, the mistake consisted of using the wrong price in the calculation; in the *Belle River* case, the mistake consisted of omitting to transfer a figure from a summary sheet to an adding-machine tape. In his judgment, Justice Southey stated in part:

[2] (1977) 15 O.R. (2d) 738
[3] Supra, at page 113

There are American cases on the point which appear to go both ways. These American cases are referred to in *Corbin on Contracts* (St. Paul; 1960), vol. 3, p. 679, where the author, after recognizing the logic of the principle followed in the Imperial Glass case, expressed the view that the result of its application is frequently unjust, depending on the circumstances of the case. I am impressed with his statement, at p. 682, that a "just and reasonable man will not insist upon profiting by the other's mistake" and that "If he does insist, it seems reasonably certain that he will get either a law suit or a poor job performed with a continuing sense of grievance." The conclusion of the learned author as to the course that should be followed by the Courts is stated at p. 688:

> Courts refusing to decree rescission for unilateral mistake often say that to do otherwise would tend greatly to destroy stability and certainty in the making of contracts. In some degree, this may be true; but certainty in the law is largely an illusion at best, and altogether too high a price may be paid in the effort to attain it. Inflexible and mechanical rules lead to their own avoidance by fiction and camouflage. A sufficient degree of stability and certainty will be maintained if the court carefully weighs the combination of factors in each case, is convinced that the substantial mistake asserted was in fact made, and gives due weight to material changes of position. Proof of the mistake should be required to be strong and convincing; but in many cases it is evident that such proof existed.

The Ontario High Court decided the *Belle River* case in favour of the contractor. The plaintiff appealed the decision to the Court of Appeal. In 1978, the Court of Appeal[4] upheld the trial judgment decision in favour of the contractor; the authorities had established a principle: an offeree cannot accept an offer that he knows has been made by mistake and that affects a fundamental term of a contract. As the Court of Appeal pointed out, in substance the purported offer, because of the mistake, was not the offer the offeror intended to make, and the offeree knew that. The situation would be quite different, the Court of Appeal noted, if the offeree had not known the offer contained a mistake, and had accepted it at face value. The court also noted that in the United States, the weight of authority is very strongly on the side of the contractor who submits a tender by mistake, or one whose tender contains a mistake when the mistake is known to the person to whom the tender is made.

[4] (1978) 20 O.R. (2d) 447

The *Belle River* case was applied in a 1979 decision of the
Ontario Court of Appeal. The case was *Ron Engineering et al.* v.
The Queen in right of Ontario et al.[5] A bid deposit cheque of
$150,000 was paid with a tender submitted to the defendant, The
Water Resources Commission. The tender concerned work to be
done for the Commission in the City of North Bay. The tender
contained a mistake similar to the one in the *Belle River* case: an
amount had been omitted from the final price. In *Ron Engineering et al.* v. *The Queen*, the omitted amount was $750,058; the
tender price was $2,748,000. The contractor was unable to con-
tact the Commission before the tenders were open. Within an
hour subsequent to the opening of the tenders, the contractor
spoke to the Commission. He also sent a telegram notifying the
Commission of the error; the telegram arrived the following morning. There was no doubt as to the genuineness of the error. The
next highest bidder had tendered a price of $3,380,464. The trial
judge found in favour of the Water Resources Commission.

The contractor appealed the decision, and the Court of Appeal
found in favour of the contractor. The court stated, in part:

> The trial Judge, as I have said, dismissed the action. In fairness
> to him, it should be pointed out that his judgment was given
> some four months before the judgment of this Court in *Belle
> River Community Arena Inc.* v. *W.J.C. Kaufmann Co. et al.*
> (1978), 20 O.R. (2d) 447, 87 D.L.R. (3d) 761, holding that an
> offeree cannot accept an offer which he knows has been made
> by mistake which affects a fundamental term of the contract.
> In our view, the principles enunciated in that case ought to be
> applied to this case. The error in question has been found to be,
> as it obviously was, material and important. It was drawn to the
> attention of the Commission almost at once after the opening
> of tenders. Notwithstanding that, the Commission proceeded
> as if the error had not been made and on the footing that it was
> entitled to treat the tender for what it said on its face.

However, the Ontario Court of Appeal's decision in *Ron Engineering* was set aside in 1981 by a decision of the Supreme Court
of Canada.[6] The decision has sparked considerable controversy
in the construction industry. The Supreme Court, focussing on
the issue of whether the contractor's tender deposit was to be

---

[5]   (1979) 24 O.R. (2d) 332
[6]   119 D.L.R. (3d) 267

forfeited, pointed out that there was a contract relating to the tender arrangements that was separate from the construction contract itself; and that the mistake in question was not communicated to the Commission at the time of tender submission and did not affect the contract relating to the tender arrangements. The following provision was contained in the Information for Tenderers:

> Except as otherwise herein provided the tenderer guarantees that if his tender is withdrawn before the Commission shall have considered the tenders or before or after he has been notified that his tender has been recommended to the Commission for acceptance or that if the Commission does not for any reason receive within the period of seven days as stipulated and as required herein, the Agreement executed by the tenderer, the Performance Bond and the Payment Bond executed by the tenderer and the surety company and the other documents required herein, the Commission may retain the tender deposit for the use of the Commission and may accept any tender, advertise for new tenders, negotiate a contract or not accept any tender as the Commission may deem advisable.

The Supreme Court held that because of the provision the contractor was contractually required to forfeit its tender deposit.

Note that the Supreme Court of Canada, in deciding *Ron Engineering*, focused on the separate contract relating to the tendering arrangements and the contractual obligation to forfeit the tender deposit. It was not prepared to accept, as the Ontario Court of Appeal had, that the mistake was relevant to the contractual obligation to forfeit the tender deposit. But note that the Supreme Court, in deciding *Ron Engineering*, did not expressly overrule the *Belle River* decision. This fuelled some speculation that future decisions might distinguish *Ron Engineering* and *Belle River* on their respective facts; however, that prospect has not yet proven of much assistance to contractors in similar cases.

One example where a trial court was prepared to distinguish between *Belle River* and *Ron Engineering* to the contractor's benefit is *Calgary* v. *Northern Construction Company Division of Morrison-Knudsen Company Inc. et al.*,[7] a 1982 case that similarly involved a clerical error of $181,000 on a tender price of $9,342,000. (It is emphasized that this example, however, is

---

[7]   (1982) 23 Alta. L.R. (2d) 388

perhaps of academic interest only, insofar as the decision of the trial court in favour of the mistaken contractor was subsequently reversed by the Alberta Court of Appeal, and ultimately the Supreme Court of Canada also decided that the case should be decided in favour of the owner.) In the *Northern Construction* case, when the contractor refused to execute the contract, the owner sued for the difference between the contractor's tender price and the second lowest bidder's price, which was $9,737,000. The Alberta Queen's Bench applied *Belle River* and distinguished the *Ron Engineering* case. In doing so the Alberta Trial court pointed out, in part:

> The similarities between the case at bar and Ron Enrg. are numerous and include:
>
> 1. The contractor's error in completing the formal tender was clerical in nature;
> 2. The error was not apparent on the face of the tender;
> 3. The error was promptly reported to the owner by the contractor;
> 4. The error was an honest one, committed unwillingly, and the conduct of the contractor was free from improper or dishonest motive;
> 5. The formal tender was irrevocable within the stated time period;
> 6. The contractor did not purport to withdraw its tender at any time.
>
> However, there are two crucial dissimilarities that distinguish Ron Enrg. and render it inapplicable to the case at bar:
>
> 1. The city of Calgary accepted the contractor's tender....
> 2. ... The parallel clause in the information to tenderers herein is cl. 8 quoted in para. 9 of the facts herein. It contains no provision similar to para. 13 in Ron Enrg....
>
> Consequently, the case at bar remains to be decided by determining the applicability of the law of mistake to the formation or enforcement of the contract arising from the acceptance by the city of the contractor's tender. The authoritative case on that issue is the Ontario Court of Appeal decision in *Belle River Community Arena Inc.* v. *W.J.C.*

Kaufman Co. Ltd ... considered by the Supreme Court in Ron Enrg. but neither over-ruled nor disapproved by that court....

As indicated, on appeal, the Alberta Court of Appeal[8] disagreed with the trial court's conclusion that the *Northern Construction* case should be decided on a basis consistent with the *Belle River* decision. Rather, the Alberta Court of Appeal applied the *Ron Engineering* rationale and held in favour of the owner. Accordingly, the contractor was caught by its clerical mistake. In its discussion, the Alberta Court of Appeal pointed out that prior to the *Ron Engineering* case an advertisement for tenders was not considered an offer but merely an invitation to tenderers to submit offers. However, *Ron Engineering* constituted authority for the principle that an advertisement for tenders constitutes an offer that is accepted when a tender is submitted and a contract is thereby formed that precludes the contractor from withdrawing its bid.

On further appeal, the Supreme Court of Canada[9] agreed that the contractor, Northern Construction, was governed by the Supreme Court of Canada's precedent decision in *Ron Engineering* and hence unable to avoid the consequences of its clerical and honest error in the circumstances of the case.

---

[8] [1986] 2 W.W.R. 426
[9] [1988] 2 W.W.R. 193

CHAPTER SIXTEEN

TENDERING ISSUES — CONTRACT A

In rendering its decision in *Ron Engineering*, the Supreme Court of Canada also confirmed a very significant (and controversial) new principle in the law of tendering in Canada. The principle is that there are two separate contracts arising in the tendering process. The first is "Contract A" (the contract of irrevocability), that deals with the tendering phase. The second is "Contract B" (the construction contract itself) that is formed on the award and that applies to the construction phase. In *Ron Engineering*, the Supreme Court of Canada held, in effect, that the owner's request for tenders constitutes an offer and that a Contract A is formed when the owner's offer is accepted upon the submission of each bid. Previously, each tender was regarded as an offer and award of contract constituted acceptance. Now, when a bid is submitted on an irrevocable basis pursuant to the conditions in the owner's request for tenders a Contract A is formed. Note that as many Contracts A will be formed as the number of tenders submitted. The second contract, the construction contract, arises on the selection of the winning bid.

The decision in *Ron Engineering* may be advantageous to an owner where a contractor has made a clerical mistake in its tender documents. However, the decision may impact to the owner's disadvantage in the contract negotiation process. The creation of a number of "Contracts A" with various bidders, each subject to the provisions of the tender package, may place the owner in a very difficult position if it chooses to negotiate with any of the bidders in a manner that is not equally applicable to all bidders. If it does so, the owner may risk breaching the other "Contracts A." As a result, sophisticated owners draft their instructions to bidders to provide for as much flexibility as possible in dealing with bidders. However, the "Contract A" concept often provides a basis for contractors to take issue and make claims should another bidder receive, for example, more advantageous opportunities to negotiate with the owner prior to the contract award. Claims arising on tendering issues are very common in Canada. It is important to understand the Contract A concept. The following cases illustrate some of the issues.

One illustrative Ontario case is *Ben Bruinsma & Sons Ltd.* v. *Chatham.*[1] In *Ben Bruinsma*, a number of tenders were submitted; accordingly, a number of "Contracts A" formed. Subsequently, the owner deleted an item from the tender package with the result that a bidder other than the original low bidder became the lowest. The original low bidder claimed damages for breach of Contract A. The tender package did not specify that items could be deleted. The court agreed that the Contract A had been breached.

Accordingly, when putting the tender package together, it is advisable to make it as flexible as possible from the owner's perspective. For example, include the right to delete items, the right to make changes, the right to overlook mistakes. You may want to go so far as to include the right to negotiate, selectively. These are some of the additional conditions that the Contract A concept has generated in instructions to bidders, beyond the advisable provision that the lowest or any tender need not necessarily be accepted. But an owner should be careful to limit the conditions in its tender package reasonably in the circumstances of each project. If an owner tries to introduce rights that are viewed as unfair by potential bidders it may find that some contractors refuse to bid or submit high prices sensing that the owner intends to take an overly aggressive contracting approach.

Another "tendering" case is *Peddlesden Ltd.* v. *Liddell Construction Ltd.,*[2] a 1981 decision of the British Columbia court. A subcontractor submitted a bid together with a bid bond for masonry work. The bid was utilized by the general contractor. The subcontractor, however, had made a mistake; it hadn't sealed its bond as required. The subcontractor was prepared to put its seal on afterwards and the bonding company was also willing to confirm the enforceability of the bond. However, the general contractor had the work performed by another subcontractor because of the missing seal. The subcontractor brought an action for damages. The court sympathized with the subcontractor and found

[1] (1984), 11 C.L.R. 37
[2] (1981), 128 D.L.R. (3d) 360

that the general contractor was obligated to use the subcontractor it had utilized in its tender to the owner. The court found that the mistake was a mere omission that could have been corrected without affecting the rights and obligations of the parties, and it found in favour of the subcontractor. The case is illustrative of both the binding nature of the bid obligations between the contractor and subcontractor and of the equitable approach that the British Columbia Supreme Court was prepared to take in the circumstances.

Mawson Gage Associates Ltd. v. *R.*[3] is a 1987 decision of the Federal Court of Canada, Trial Division. A subcontractor prepared an estimate on the basis of plans received directly from the owner. The subcontractor's bid was filed with the general contract. However, several pages of specifications for the project were missing from the tender details provided to the subcontractor. The omission of the pages was the owner's error. The general contractor and the owner were notified of the mistake. The subcontractor repeatedly attempted to resolve the problem but was unable to do so. The subcontractor signed and completed the work. The subcontractor then sued the owner in tort for losses arising from negligent misrepresentation. The subcontractor succeeded. Otherwise, as the court pointed out, the owner would have been unjustly enriched. *Mawson Gage* illustrates the importance of close care and attention when compiling specifications.

A case that involved "last minute" telephone bids from subcontractors (a common practice) is *Gloge Heating and Plumbing Ltd.* v. *Northern Construction Company Ltd.*,[4] a 1986 Alberta Court of Appeal decision. The subcontractor had telephoned its tender to the contractor minutes before tenders closed. This practice is used to avoid bid shopping by general contractors. The general contractor included the subcontractor's tender and filed its own tender. The subcontractor had made a serious mathematical error. The subcontractor advised the contractor of the error. The owner refused to allow the contractor to adjust its tender. The subcontractor refused to perform. Another subcontractor performed for an extra $340,000. The general contractor successfully sued the original subcontractor. The court confirmed that telephone bids are just as binding as written tenders. The subcontractor could only have withdrawn its bid prior to the close of the general contract bidding. The subcontractor's bid was irrevocable

---

[3]   (1987), 13 F.T.R. 18
[4]   27 D.L.R. (4th) 264

for the same time period as the general contractor's, not surprisingly given the Contract A concept.

Forest Contract Management Ltd. v. *C&M Elevator Ltd.*[5] also illustrates the similar binding nature of subcontract bids. The contractor requested a quote from a subcontractor and submitted a lump sum proposal. The contract between the owner and the contractor was finalized. The subcontractor had omitted one elevator in its estimate. The contractor called the owner and attempted to withdraw. The owner retendered and claimed against the contractor for breach of contract and for the difference between the original and newly accepted bid. The court found that the contractor was in breach of the contract and the owner was awarded damages. That is consistent with the Contract A approach and the enforceability of the conditions in the tender package.

Westgage Mechanical Contractors Ltd. v. *PCL Construction Ltd.*[6] illustrates that a subcontractor can be disadvantaged where the owner and the general contractor negotiate different terms prior to awarding the contract to the general contractor. In *Westgage*, negotiations occurred between the owner and the lowest three tenderers. However, the subcontractor was not consulted. As a result of the negotiations, the terms of the contract changed. This resulted in a reduced price and a reduced completion time. The subcontractor was not consulted during the negotiations and it refused to alter its original estimate. When the work was awarded to a different subcontractor, the original subcontractor sued. The court pointed out that the renegotiation of the contract terms amounted to a counter offer that released the general contractor's commitment to the original subcontractor.

On the issue of whether or not there is a duty to accept the lowest bidder, some provincial Canadian courts have differed on the outcome. Final resolution of the issue awaits a decision by the Supreme Court of Canada.

For example, in *Chinook Aggregates Ltd.* v. *Abbotsford (Municipal District)*,[7] a 1989 decision of the British Columbia Court of Appeal, the municipality invited bids on a gravel crushing contract. The municipality had an unstated policy of preferring local contractors. The British Columbia Court of Appeal held that the provision that "the lowest or any tender will not necessarily

---

[5]   (1988), 93 A.R. 38
[6]   (1987), 25 C.L.R. 96
[7]   35 C.L.R. 242

be accepted" did not override the implied duty to accept the lowest bidder. The court found the owner had to adhere to the established custom of accepting the lowest bid unless the tender documents expressly stated that the bids would be evaluated on other criteria.

Ontario courts have not followed the British Columbia position. *Megatech Contracting Ltd.* v. *Regional Municipality of Carleton*,[8] is one such Ontario case. The plaintiff's bid was the second lowest. The lowest bid failed to follow the tender guidelines. It omitted to name certain proposed subcontractors. The lowest bidder was awarded the contract and the plaintiff sued for breach of contract. The court dismissed the action stating that the Region had the flexibility, as set out in the tender package, to award the contract to the lowest or any tenderer.

Acme Building & Construction Ltd. v. *Newcastle (Town)*[9] is a similar 1992 decision of the Ontario Court of Appeal. Although Acme was the low bidder for a new municipal centre in Newcastle, the next lowest bidder had specified a much earlier completion date and nominated local subcontractors for more of the work; 23 percent compared to 18 percent. The Town Council concluded that the earlier completion date would save them $25,000 in rent, more than the difference in the tender prices. Council also preferred the higher bid because more local subcontractors would work on the project. Acme argued that such consideration should have been made clear and that custom and usage should prevail. The court said that custom and usage cannot prevail over the express language in the tender documents. The tender documents had provided that the owner would have the right not to accept the lowest or any other tender.

A Saskatchewan case of interest is *Kencor Holdings Ltd.* v. *Saskatchewan*.[10] In *Kencor Holdings* the tender documentation included a clause allowing the Minister of Transport to "refuse to accept any tender, waive defects or technicalities, or ... accept any tenders that he considers to be in the best interests of the Province." The court found that the clause was too vague to be meaningful. The court did not allow the government to rely on the clause to justify rejecting the lowest tender. The court felt that the government should be forced to state the criteria upon which it bases its decisions, as a check on the arbitrary use of

[8] (1989), 68 O.R. (2d) 503
[9] (1992), 2 C.L.R. (2d) 308
[10] [1991] 6 W.W.R. 717 (Sask. Q.B.)

power. The court found the government liable for the lowest bidder's loss of profit.

A 1994 decision of the Alberta courts emphasizes the importance of the provisions in the invitation to tender and the owner's ability to act accordingly. The case is *North American Construction Ltd.* v. *City of Fort McMurray.*[11] The invitation to tender contained the following provision: "The owner reserves the right to reject any or all tenders or to accept the tender deemed most favourable in the interests of the City of Fort McMurray." Accompanying the invitation, the instructions to bidders provided: "The owner reserves the right to accept or reject any or all tenders and to waive irregularities and formalities at his discretion. The lowest tender will not necessarily be accepted ...."

Three bids were submitted. The contract was awarded to a bidder whose price was higher than the plaintiff's bid. The plaintiff sued the City of Fort McMurray for breach of Contract A. The court found in favour of the City, pointing out that it is not an implied term of the invitation to tender that the contract will be awarded to the lowest qualified bidder. As the court pointed out: "there cannot be an implied term which conflicts with an express term ...."

The court in *North American Construction*, in the decision, included the following practical perspective: "The words used by the Defendant in the invitation to tender and instructions to bidders are short, old, simple and readily understandable. They must not be deliberately misinterpreted by the courts to arrive at a result-oriented decision based on some vague notion of "fairness." Some courts appear determined to make business life as legally difficult as possible for businessmen. I am not one of those. As a wise old saying goes — if it ain't broke don't fix it."

Bid protests have become a very common occurrence in the Canadian construction industry. The Contract A concept has provided opportunities for creative challenges. Close attention to the provisions of Contract A in putting together the wording of the tender package remains an important and advisable objective.

---

[11] 16 C.L.R. (2d) 225

CHAPTER SEVENTEEN

CONTRACT INTERPRETATION

Parties to a contract sometimes dispute the meaning of part of the contract. Such disputes can be referred to the court. The court examines the specific wording of the part of the contract in question; and interprets the contract to determine its most reasonable meaning. The court may refer to dictionary definitions and to the intent of the parties who have entered into the contract. The parties to the contract will be bound by the court's determination of the most reasonable interpretation.

In approaching the interpretation of contracts there are different approaches that can be taken — the "liberal" approach or the "strict" construction approach. The "liberal" approach takes into account the intent of the parties and, in the extreme, may lead to too much speculation on that intent. The "strict" approach focuses on the precise words in the agreement, in the extreme relying on dictionary meanings. Obviously problems can result from too rigorous an application of either approach. In the end, it will be the court's decision as to what constitutes the most reasonable interpretation in the circumstances. The two approaches, however, point to the importance of clear and careful contract wording.

The approach of the Supreme Court of Canada in its 1989 decision in *Hunter Engineering Company* v. *Syncrude Canada Ltd.*[1] emphasizes the Supreme Court's endorsement of the approach that favours a strict construction on contract wording, provided the contract is not "unconscionable" (an unlikely possibility where it is between business parties). It is interesting, however, that not all of the Supreme Court Justices would have adopted the strict construction approach that the majority of the court agreed upon in the *Syncrude* v. *Hunter* case. Two of the Justices

[1] [1989] 1 S.C.R. 426

favoured the approach of considering whether, in all the circumstances, it would be "fair and reasonable" to enforce a contract clause.

In interpreting the contract, the court may listen to witnesses. The witnesses may testify as to the intention of the persons who signed the contract; they may also discuss the contract's subject matter. The court may be required to judge the relative credibility of the witnesses. It is obviously preferable to avoid the problems involved in determining contract meaning by ensuring that contracts are prepared with sufficient clarity in order that the likelihood of the need for court interpretation will be minimized.

Engineering projects and business transactions, in Canada and elsewhere, involve important contract documents and appropriate legal advice should be obtained in the interests of protecting both the client and the engineer in the negotiation and preparation of contract documents.

RULE OF *CONTRA PROFERENTEM*

A contract rule of interpretation that underscores the importance of clear and unambiguous language in the drafting of contracts is the "rule of *contra proferentem*." Simply put, the rule provides that where a contract is ambiguous, it will be construed or interpreted against the party that drafted the provision. Accordingly, it is most important to avoid ambiguities. The rule of *contra proferentem* provides a convenient basis upon which to attack a poorly drafted and ambiguous document.

PAROL EVIDENCE RULE

A contract should embody all terms agreed upon by both parties. Problems can occur when terms that are agreed upon verbally are not included in the written contract that embodies the agreement. If a condition is agreed upon verbally but is not included in the contract, the condition is not part of the contract. The contract law rule that precludes evidence of the omitted condition is called the "parol evidence rule." It is most important that contracts be carefully drafted and that no agreed-upon conditions are omitted from the final contract form.

There are, however, exceptional circumstances in which courts will not always apply the parol evidence rule. For example, where it can be substantiated that a contract was to be effective only if an agreed-upon condition were to occur. The court considered such a situation in *Pym* v. *Campbell*.[2] A contract concerning shared ownership of the invention of a machine was entered into. It was established at trial that, during the financial negotiations, the parties had agreed that the purchase of the invention rights would be conditional: the invention would have to be approved by two engineers. There was no mention of this condition in the written contract. Two engineers were approached, but only one expressed a favourable opinion of the invention. The second engineer refused to do so. The defendants contended that the condition precedent had not been met, and that therefore no contract was formed. The court accepted the defendants' argument, and admitted evidence of the condition precedent. In part, the court stated:

> No addition to or variation from the terms of a written contract can be made by parol: but in this case the defence was that there never was any agreement entered into. Evidence to that effect was admissible; and the evidence given in this case was overwhelming. It was proved in the most satisfactory manner that before the paper was signed it was explained to the plaintiff that the defendants did not intend the paper to be an agreement till Abernethie had been consulted ...

The "parol" evidence rule relates to evidence extrinsic to or "outside" the written contract prior to the execution of the contract. Only in exceptional circumstances is a contract affected by extrinsic evidence. But parties to a contract are free at any time to alter the terms of the contract after it has been signed, provided both parties agree and provided the essential contract elements are present in the amended contract.

IMPLIED TERMS

Occasionally, the parties to a contract overlook the inclusion of an obvious term. Where it is clearly reasonable to do so, the courts may give business efficacy to an agreement through "implication of terms." A leading case example is known as "*the Moorcock*."[3] The plaintiff was the owner of a steamship called

---

[2]   [1856] 119 E.R. 903
[3]   [1889] 14 P.D. 64

the *Moorcock*. He paid for space at the defendants' wharf and jetty on the Thames. While the *Moorcock* was docked, the tide went out; the *Moorcock* settled on a ridge of hard ground and was damaged. The court held that the parties must have intended that the vessel would have been safe at low tide and hence implied a term accordingly in finding the defendants liable.

A more recent example is the case of *Markland Associates Ltd.* v. *Lohnes*,[4] a 1973 decision of the Nova Scotia Supreme Court, Trial Division. The court held, in the absence of express terms, that the building contract in question implied several things: that the materials and workmanship should be of a proper standard or quality; that the work was to be carried out in a proper and workmanlike manner; that the work and materials, when completed and installed, would be fit for the purposes intended; and that the work would be completed in a reasonable time and without undue delay.

In 1961, the Ontario Court of Appeal heard *Pigott Construction Co. Ltd.* v. *W.J. Crowe Ltd.*[5] The court had to decide the advisability of implying a term into a contract. The court referred to earlier decisions involving implied terms in summarizing the principle:

> I have for a long time understood that rule to be that the Court has no right to imply in a written contract any such stipulation, unless, on considering the terms of the contract in a reasonable and business manner, an implication necessarily arises that the parties must have intended that the suggested stipulation should exist.

In 1983, the Ontario Court of Appeal heard *G. Ford Homes Ltd.* v. *Draft Masonry (York) Co. Ltd.*,[6] a case involving the supply and installation of two circular staircases. The court decided the contract implied that the staircases would comply with the requirements of the Ontario Building Code. The subcontractor, G. Ford Homes Ltd., a fabricator and installer of residential staircases, was engaged to provide the staircases. Its representative attended at the homes in question, but declined an opportunity to review the architectural plans, which indicated the requisite clearance to comply with the Code. Instead, he took

[4]   [1973] 33 D.L.R. (3d) 493
[5]   27 D.L.R. (2d) 258
[6]   43 O.R. (2d) 401

some measurements and offered a selection of staircases to the contractor. The installed stairways did not comply with the Ontario Building Code because the head room was one and a half inches short of the specified minimum. The building inspector said the staircases had to be replaced; and the subcontractor commenced a lawsuit to recover the cost of supplying and installing the stairs. The Ontario Court of Appeal pointed out that the subcontractor was an "expert" in the manufacture and installation of stairs, and was fully aware of the requirements of the building code. The court also said it was natural and reasonable, in the circumstances of the case, to rely upon the subcontractor to supply and install the staircases in compliance with the Ontario Building Code, and that it would be unrealistic to come to any other conclusion. The court pointed out that there could be no business efficacy to the contract without implying such a term, and that to sanction the installation of such a staircase in contravention of the Code would be tantamount to sanctioning an illegal contract. The principle enunciated in 1889 in the *Moorcock* case was applied.

STUDY QUESTIONS

The following questions are included for study purposes:

1. What is the difference between the liberal and strict construction approaches to interpreting contracts?

2. Explain the rule of *contra proferentem*. How might it be significant as far as engineering drawings and specifications are concerned?

3. It is advisable to put contracts in writing. Explain how the parol evidence rule can affect a written contract.

4. A contract may contain both "express terms" and "implied terms." What is the difference between such terms? When will a court be likely to make use of implied terms in interpreting a contract? Give an example of an implied term that could be included in a construction contract.

CHAPTER EIGHTEEN

DISCHARGE OF CONTRACTS

There are several ways to accomplish the discharge of a contract.

PERFORMANCE AS A MEANS OF DISCHARGE

When all parties to a contract have completed their respective obligations, the contract is at an end. So long as any obligations described in the contract remain unfulfilled, the contract remains in effect.

Some contracts provide that obligations will continue beyond initial performance and payment for services and materials. An example is an equipment supply contract that contains warranty provisions; the manufacturer undertakes to remedy defects within a specified time period. In building contracts, the contractor normally undertakes similar warranty obligations.

AGREEMENT TO DISCHARGE

The parties to a contract are always free to amend the contract; thus they can subsequently agree to cancel or terminate the contract upon mutually agreeable terms and conditions.

DISCHARGE PURSUANT TO EXPRESS TERMS

It is advisable to include, in a contract, provisions whereby any or all parties may terminate the contract upon the occurrence of certain events. For example, a contract might terminate upon the bankruptcy of one of the parties. Construction contracts often provide for termination if an engineer determines that the contractor has failed to complete the work properly, or has otherwise failed to substantially comply with the requirements of the contract.

DISCHARGE BY FRUSTRATION

At times, without default by either party to a contract, changing circumstances may radically change the obligations of the parties. If this happens, the contract will have been "frustrated," and is discharged by such frustration. However, the doctrine of frustration may not be used to justify discharge of a contract simply because circumstances have made performance more onerous than contemplated. Such an application of the doctrine would be contrary to the general principles that support the binding effect of contracts. The doctrine of frustration will be applied by a court only where exceptional circumstances, which were not contemplated by the parties, have arisen, and only where discharge by frustration is the only practical and reasonable solution.

For example, in 1917 the English House of Lords heard the case of *Metropolitan Water Board* v. *Dick, Kerr and Company, Limited.*[1] A contract had been entered into in July of 1914 for the construction of a reservoir over a six-year period. But because of special wartime legislation, the contractor was ordered to cease work in 1916 by the Ministry of Munitions. The character and duration of the wartime interruption changed the contract; when the contract was resumed, it was a different contract than that which had been entered into. The House of Lords held that the contract had been discharged by frustration, in the circumstances.

Equipment supply agreements and construction contracts often contain a *"force majeure"* provision. The *force majeure* clause usually provides that time for completion will be extended in the event of war, riot, insurrection, flood, labour dispute, or other events that arise beyond the control of either party. The contract being discussed in *Metropolitan Water Board* v. *Dick, Kerr and Company, Limited* did contain such a *force majeure* provision. The Water Board argued that the cease-work order should have been dealt with by an extension of time, according to the provision. The House of Lords did not apply the *force majeure* provision, but rather emphasized the very exceptional wartime circumstances affecting the contract. The case illustrates the very important principle that the doctrine of frustration will be applied to discharge a contract only in most exceptional circumstances.

[1]   [1918] A.C. 119

In 1956, the House of Lords heard *Davis Contractors Ltd.* v. *Fareham Urban District Council.*[2] Completion of a building contract had been delayed due to the scarcity of labour. Neither party was in default. The contract required the contractor to build seventy-eight houses within an eight-month period. Because of the labour shortage, twenty-two months were needed to complete construction. The House of Lords did not accept the argument that the contract had been frustrated, concluding that there had been an unexpected turn of events, which simply rendered the performance more onerous than contemplated. The House of Lords pointed out:

> But, even so, it is not hardship or inconvenience or material loss itself which calls the principle of frustration into play. There must be as well such a change in the significance of the obligation that the thing undertaken would, if performed, be a different thing from that contracted for.
>
> Two things seem to me to prevent the application of the principle of frustration to this case. One is that the cause of delay was not any new state of things which the parties could not reasonably be thought to have foreseen. On the contrary, the possibility of enough labour and materials not being available was before their eyes and could have been the subject of special contractual stipulation. It was not made so. The other thing is that, though timely completion was no doubt important to both sides, it is not right to treat the possibility of delay as having the same significance for each. The owner draws up his conditions in detail, specifies the time within which he requires completion, protects himself both by a penalty clause for time exceeded and by calling for the deposit of a guarantee bond and offers a certain measure of security to a contractor by his escalator clause with regard to wages and prices. In the light of these conditions the contractor makes his tender, and the tender must necessarily take into account the margin of profit that he hopes to obtain upon his adventure and in that any appropriate allowance for the obvious risks of delay. To my mind, it is useless to pretend that the contractor is not at risk if delay does occur, even serious delay. And I think it a misuse of legal terms to call in frustration to get him out of his unfortunate predicament.

[2] [1956] A.C. 696

A similar decision was reached in 1963, when the Manitoba Court of Appeal heard *Swanson Construction Company Ltd.* v. *Government of Manitoba; Dominion Structural Steel Ltd., Third Party*.[3] A contractor was forced to work in winter conditions rather than summer conditions, as planned, because the work site was not available soon enough. The contractor argued frustration of contract. But the court refused to invoke the doctrine of frustration. The court pointed out that in many building contracts, some delay in performance may occur; the contractor should have adjusted his tender price accordingly.

SAMPLE CASE STUDY

The following hypothetical case and commentary is included for illustrative study purposes.

1. A construction company entered into a construction contract to build 45 houses at a mine site in northern Ontario within a twelve-month period. At the end of the twelve-month period, only 30 houses had been completed and the company abandoned the project claiming that the local labour shortage had made it impossible for it to perform pursuant to the terms of the contract. The company emphasized that it had never contemplated the labour shortage when it had provided its quotation for the project to the mine site owner. The company claimed that the contract had been frustrated. Was the company justified, from a legal point of view, in abandoning the project? Explain.

Commentary: 1. This is obviously very similar to the *Davis Contractors* case described in the chapter. In answering, explain when frustration would be effective to discharge a contract and compare with the facts given here. A simple example of a scenario that may be helpful to explaining frustration is when a renovation contract cannot be performed because the building has burned to the ground.

---

[3]   40 D.L.R. (2d) 162

CHAPTER NINETEEN

BREACH OF CONTRACT

If a party to a contract fails to perform obligations specified in the contract, then the defaulting party has breached the contract. The innocent party is entitled to certain remedies; the particular remedy will depend upon the nature of the breach and the terms of the contract. For example, a breach of contract may entitle the non-defaulting party to sue for damages sustained as a result of the breach. The non-defaulting party may also be entitled to regard the contract as discharged because of the breach.

An obligation essential or vital to the contract is called a "condition"; an obligation that is not essential to the contract is called a "warranty." Breach of a condition or of a warranty may entitle the non-defaulting party to damages. But only breach of a condition that is of fundamental importance to the contract will entitle the non-defaulting party to consider the contract discharged by the breach.

Note, however, that the term "warranty" has several meanings. It can be used to describe a minor term of a contract; at other times, it can be used to mean "guarantee." For example, a manufacturer could guarantee the performance of his equipment, by issuing a warranty. The warranty issued may be an essential term of the equipment supply contract. Establishing whether a provision of a contract is a condition or a warranty may be a key issue in a law suit. In the case of *Piggot Construction Co. Ltd.* v. *W.J. Crowe Ltd.*,[1] the Ontario Court of Appeal quoted from the *Law of Contract*, Cheshire & Fifoot, 5th edition, page 488:

> Breach, no matter what form it may take, always entitles the innocent party to maintain an action for damages, but it does not always discharge the contract.... when can a breach be regarded as a cause of discharge? The manner in which from

[1] 27 D.L.R. (2d) 258

time to time the answer to this question has been judicially expressed has not been altogether uniform, and it has been clouded by a distinction between dependent and independent promises that appeared of greater importance in the past than it does in the modern law. But at the present day it is possible to state with some confidence what the position is.

A breach of contract is a cause of discharge only if its effect is to render it purposeless for the innocent party to proceed further with performance. Further performance is rendered purposeless if one party either shows an intention no longer to be bound by the contract or breaks a stipulation of major importance to the contract....

It may, indeed, be said in general that any breach which prevents substantial performance is a cause of discharge. Whether performance is substantially prevented or only partially affected is, of course, a question that depends upon the circumstances of each case.

In the *Pigott Construction* case, the contractor was under an obligation to proceed as expeditiously as possible; he was also obligated to provide temporary heat in the buildings during winter construction. The Court of Appeal held that neither obligation could be regarded as fundamental; neither obligation affected the substance and foundation of the transaction between the parties. Hence a breach of either provision could not be regarded as sufficient cause to discharge the contract.

It is often difficult to establish whether a breach might entitle the non-defaulting party to treat the contract as terminated. To avoid the difficulty, construction contracts very often contain a special provision: if the engineer determines that the contractor's performance has been inadequate, then the contract may be terminated by the engineer's client, the owner.

REPUDIATION

When one party to a contract expressly tells the other party that he or she has no intention of performing contractual obligations, the declaring party has repudiated the contract. The defaulting party need not express his or her intentions verbally; but might indicate by conduct that the contractual obligations will not be performed. The non-defaulting party can either ignore the breach — in which case the contract continues — or can assume that the contract has been discharged by the repudiation. If the non-defaulting party treats the contract as discharged, he or she may claim damages against the defaulting party. The right to elect to discharge the contract makes it impossible for a defaulting party to avoid contractual obligations by announcing that he or she has no intention of fulfilling the contract. If the non-defaulting party elects to discharge the contract, the non-defaulting party is required to communicate his or her intention to the defaulting party "with reasonable dispatch."

REMEDIES

A non-defaulting party is entitled to damages for losses incurred as a result of breach of contract. The injured party may also be entitled to a *quantum meruit* remedy; he or she may also be eligible for equitable remedies called "specific performance" and "injunction."

The court must determine the amount of damages to be awarded as a result of a breach of contract, by applying long-established principles. As stated in the landmark 1854 English case decision in *Hadley* v. *Baxendale*:[2]

> Where two parties have made a contract which one of them has broken, the damages which the other party ought to receive in respect of such breach of contract should be such as may fairly and reasonably be considered either arising naturally, i.e., according to the usual course of things, from such breach of contract itself, or such as may reasonably be supposed to have been in the contemplation of both parties, at the time they made the contract, as the probable result of the breach of it.

Hence, damages should flow naturally from the breach or be reasonably foreseeable by both parties at the time of entering into the contract. If the contract were entered into under special circumstances, and if those special circumstances were communicated between the parties at the time when the contract was formed, then those special circumstances would be taken into account in determining the damages resulting from the breach. The *Hadley* v. *Baxendale* judgment is of further assistance:

---

[2] (1854) 9 Exch. 341

Now, if the special circumstances under which the contract was actually made were communicated by the plaintiffs to the defendants, and thus known to both parties, the damages resulting from the breach of such a contract, which they would reasonably contemplate, would be the amount of injury which would ordinarily follow from a breach of contract under those special circumstances so known and communicated. But, on the other hand, if these special circumstances were wholly unknown to the party breaking the contract, he, at the most, could only be supposed to have had in his contemplation the amount of injury which would arise generally, and in the great multitude of cases not affected by any special circumstances, from such a breach of contract....

The plaintiffs in the *Hadley* v. *Baxendale* case operated a mill. The plaintiffs asked the defendants, who were carriers, to deliver a broken crank shaft to its manufacturer for repairs. Through the defendants' neglect, delivery of the shaft was delayed. The crank shaft was essential to the mill's operation, but the plaintiffs did not communicate its importance to the defendants. The plaintiffs brought an action for damages for lost profit during the delay period. As the court noted, however, the defendants were not told that lost profits would result from a delay in the delivery of the shaft.

DIRECT AND INDIRECT DAMAGES

A distinction should be drawn between direct and indirect damages.

Direct damages are perhaps best explained by the following illustration. An owner receives bids from three contractors on a project. The owner awards the contract to the lowest bidder. The lowest bidder refuses to perform and thereby immediately defaults. The owner then awards the contract to the next lowest bidder. The direct damages the owner has suffered are equal to the amount by which the price of the second lowest bidder exceeds the lowest bid. The lowest bid was the contract price the owner had originally agreed upon and for which the work should have been performed. Note that the damages are equal to the extra cost beyond the original contract price.

Indirect damages, on the other hand, are perhaps best illustrated as "consequential" to the breach and might include damages for "lost profits" caused by a plant shut-down resulting from a contractor failing to perform services properly and thereby affecting the overall operation of the plant. An example might be where an outside contractor working within a factory cuts a power line that interferes with the operation of the plant, a grim prospect for all concerned. Another example of an indirect damage is a fine that might be levied against a plant owner as a result of a contractor's non-compliance with an environmental protection statute. Indirect damages are often referred to as "special, indirect, or consequential damages."

Today, contracting parties often seek to limit the extent of damages for which they might be responsible. Liability may be limited in the terms of the contract itself. For example, an equipment supply contract might advisedly state that:

"In no event whatsoever will the manufacturer be responsible for any indirect or consequential damages howsoever caused."

Such a limitation provision is certainly advisable from the supplier's perspective, given the *Hadley* v. *Baxendale* decision that confirmed that damages for breach of contract should be those reasonably contemplated on the basis of special circumstances known at the time of contracting. In order to avoid facing creative interpretations in that regard, it is far preferable and not unreasonable from a supplier's perspective to expressly preclude indirect or consequential damage claims by including an appropriate exemption provision whenever possible.

DUTY TO MITIGATE

A party that suffers a loss through a breach of contract must take reasonable steps to mitigate or reduce the amount of damages suffered. The plaintiff is expected to behave in a reasonable manner in mitigating damages. If the plaintiff does not, that conduct will be taken into account when the court is fixing the damage award.

PENALTY CLAUSES

Contracts often contain provisions whereby a party is required to pay prescribed damages if a certain event occurs — for example, if the contract is not completed by a specified date. However, the parties must make a genuine attempt at the time of entering into the contract to pre-estimate the amount of damages likely to occur as a result of such breach; otherwise, the court will not uphold such provisions. In contracts these pre-estimated damages are called "liquidated damages," and are very common on equipment supply and construction contracts. As previously noted, the amount of damages awarded by a court will be based upon the actual damages that result from the breach and in contemplation of (or foreseeable to) the parties at the time of the formation of the contract. Hence a court will not enforce a penalty clause that does not represent a genuine pre-estimate of damages. The use of the term "penalty" should be avoided in contracts. Use the term "liquidated damages" rather than "penalty." Avoid the connotation of penalizing the contractor.

QUANTUM MERUIT

Suppose that certain services have been requested and performed, but that no express agreement was reached between the parties as to what payment would be provided in return for the services. In such a situation, the court will award payment by implying that the party performing the services ought to be paid a reasonable amount, that is, an amount determined on the basis of *quantum meruit* — "as much as is reasonably deserved" for the time spent and materials supplied.

Quantum meruit may apply in other situations. For example, a contract might expressly provide for payment, but if the party obligated to pay repudiates the contract and the innocent party elects to treat it as discharged, *quantum meruit* may apply. For example in 1968, the Supreme Court of Canada decided *Alkok* v. *Grymek*.[3] The defendant owner had repudiated a construction contract with the plaintiff. The contract provided that the owner would pay the contractor a fixed amount, in accordance with an agreed-upon schedule, for completion of various parts of the work "upon the Architect's certificate (when the Architect is satisfied that payments due to Sub-Contractors have been made)." The contractor failed to satisfy the architect that sub-contractors had been paid. As well, the owner and the architect complained that there were defects in the construction, which had been delayed. The owner terminated the contract and engaged other contractors to complete the building. The court determined that the contractor's failure to satisfy the architect that sub-contractors had been

[3]   [1968] S.C.R. 452

paid did not amount to breach of an essential term of the contract. The breach did not justify the owner's termination of the contract. Nor did the evidence of defective workmanship or delay go to the root of the contract. The courts decided that the owner's repudiation of the contract was not appropriate. The contractor succeeded in his claim to recover for work done on a *quantum meruit* basis.

SUBSTANTIAL COMPLIANCE

A contractor might substantially comply with the terms of a contract, yet fail to comply with some minor aspect of the contract's provisions. The contractor will be entitled to be paid the contract price less the cost of damages caused by any such failure. This principle is called the doctrine of substantial compliance. For the doctrine to apply, however, the facts must substantiate that the contract deficiencies are of a minor nature. An example is the 1951 decision of the Ontario Court of Appeal in *Fairbanks Soap Co. Ltd.* v. *Sheppard*.[4] The court had to decide whether the contractor was entitled to payment for equipment supplied and installed. The contractor had substantially complied with the terms of the contract; there were minor defects, which were remediable without excessive cost. The Court of Appeal referred to the applicable principles as follows:

> On the question of substantial compliance, the case of *H. Dakin & Co.* v. *Lee*, [1916] 1 K.B. 566, is a leading case. The law was there laid down by Ridley J. in the Divisional Court — and the judgment of the Divisional Court was sustained on appeal — in the following language [p. 569]: "It seems to me, however, from the authorities that where a building or repairing contract has been substantially completed, although not absolutely, the person who gets the benefit of the work which has been done under the contract must pay for that benefit...."
>
> The principle stated in *Dakin* v. *Lee*, supra, is simply this, that the person who has done the work and/or supplied the materials should be paid what he deserves for what he has done. Here the trial Judge has arrived at an amount of which he concluded the defendant was deserving, by deducting the sum of $600 as the cost of making the necessary adjustments

[4]   [1952] 1 D.L.R. 417

and adding the necessary parts to put the machine in proper running order.

SPECIFIC PERFORMANCE AND INJUNCTION

The remedies of specific performance and injunction are equitable remedies; they supplement the remedy of damages. The courts will not grant the remedies of specific performance or injunction where damages provide sufficient relief.

SPECIFIC PERFORMANCE

To remedy a contract dispute, the courts may, where appropriate, require a party to a contract to perform a contractual obligation. This remedy is called "specific performance." It is most often granted in cases concerning contracts for the sale of land. The courts presume that, when one has contracted to purchase particular land and the contract is breached by the vendor, a damage award of money will not be satisfactory. Therefore the vendor will be required to convey the land, in accordance with the agreement of purchase and sale. A breach of contract for the sale of a unique item of personal property — for example, an antique automobile — may also result in a court award of specific performance. The vendor will be required to fulfil the obligation to sell the item to the purchaser only where the property in question is sufficiently unique that damages for breach of contract will clearly not provide an adequate remedy to the purchaser.

Where it would have to supervise the performance of an obligation, the court will not grant the remedy of specific performance. Hence, breach of a contract for engineering services, construction, manufacturing, or installation of machinery will not result in an award of specific performance.

INJUNCTION

An injunction is a court order that prohibits or restrains a party from the performance of some act, such as a breach of contract. A court will not grant the remedy of injunction unless the contract contains a "negative covenant"; a negative covenant is a promise not to do something. In a non-competition agreement, for example, the promise not to compete for a specific period of time within a defined geographical area is a negative covenant. A court might order an injunction to restrain a party from breaching the covenant. Before an injunction will be granted, however, the court will apply tests of reasonableness as to time and geographic limitations, and will consider public policy with respect to such "restraint of trade" contracts.

STUDY QUESTIONS

The following questions are included for study purposes:

1. Briefly describe the basis upon which damages for breach of contract are calculated at common law. What is the difference, if any, between the basis for calculating damages for breach of contract and for tort?

Commentary: 1. Point out that, in tort, damages are generally the injuries that flow from the negligence, i.e., the actual damages caused. By contrast, in contract, damages are what is reasonably contemplated or agreed at the time the contract is formed. Accordingly, in contract, an important opportunity arises to limit liabilities between contracting parties.

2. Some construction contracts contain a provision that failure of the contractor to complete the work by a specific date will result in the contractor being required to make a specified payment to the owner for each day, week, or month that completion of construction is delayed. Is such a penalty provision always enforceable? Discuss.

Commentary: 2. Point out the need for a genuine pre-estimate.

3. Briefly describe the basis upon which damages for breach of contract are calculated at common law. Explain also the meaning of "mitigating damages."

Commentary: 3. Use the *Hadley* v. *Baxendale* case as a basis for your answer concerning damages. Emphasize the importance of taking reasonable steps to mitigate and why.

CHAPTER TWENTY

FUNDAMENTAL BREACH

The doctrine of fundamental breach has received considerable attention by our courts, particularly in the past ten years. As the case illustrations will indicate, the doctrine may be very significant to engineers.

The doctrine of fundamental breach may be applied to a contract that contains an exemption clause; essentially, it renders the exemption clause ineffective in the event of a fundamental breach of contract. An "exemption clause" is a provision whereby contracting parties may limit the extent, in whole or in part, of liability that arises as a result of breach of contract.

An example is the case *Harbutt's Plasticine Ltd.* v. *Wayne Tank and Pump Co. Ltd.*[1] A contract was entered into for the design and installation of storage tanks for stearine, a greasy wax that is one of the main ingredients of plasticine. As part of the contract, the contractors designed a plastic pipeline wrapped with electrical heating tape; the pipeline was to be used to liquefy the stearine, in order to convey it from one point to another. The plastic pipe became distorted under the heat. It sagged and cracked, and the stearine escaped and became ignited. The plaintiff's factory was completely gutted by the fire. The trial judge concluded that the contractor was in fundamental breach of contract. The court stated:

> In breach of their contract the defendants designed, supplied and erected a system which was thoroughly — I need not abstain from saying *wholly* — unsuitable for its purpose, incapable of carrying it out unless drastically altered, and certain to result not only in its own destruction but in considerable further destruction and damage ... the supply of the useless and dangerous durapipe, coupled with the useless

---

[1] [1970] 1 All E.R. 225

thermostat was a breach of the basic purpose which might be described as total, going to the root of the contract.

The contract contained a provision that limited the contractor's liability for accidents and damage to £2,300. The Court of Appeal held that, because of the fundamental breach, the contractors were not entitled to rely on the liability-limiting provision. The contractors were held liable for the cost of reinstating the factory, an amount determined at trial to be in excess of £170,000.

The *Harbutt's Plasticine* case provided an important principle: in the event of a fundamental breach, that is a breach of such a nature as to go to the very root of the contract, an exemption clause in a contract would not afford protection to the party that committed the fundamental breach. But readers should be aware of a major shift away from that principle that has occurred in our courts.

The *Harbutt's Plasticine* precedent has been applied by Canadian courts, particularly between 1970 and 1980. However, the precedent was dramatically overruled in England in a 1980 decision by the English House of Lords, in a case called *Photo Production Ltd.* v. *Securicor Transport Ltd.*[2] The House of Lords stated that the whole foundation of the *Harbutt's Plasticine* case was unsound.

In *Photo Production*, a security contract was entered into between a manufacturer and a security company. During a night patrol at the factory, one of the employees of the security company started a fire. The fire spread out of control and destroyed the factory and its contents, together valued at £615,000. The contract contained an exemption clause that limited the contractor's liability:

> Under no circumstances shall the Company [Securicor] be responsible for any injurious act or default by any employee of the Company unless such act or default could have been foreseen and avoided by the exercise of due diligence on the part of the Company as his employer; nor, in any event, shall the Company be held responsible for; (a) Any loss suffered by the customer through burglary, theft, fire or any other cause, except insofar as such loss is solely attributable to the negligence of the Company's employees acting within the course of the employment....

[2]   [1980] 1 All E.R. 556

No negligence was alleged in the *Photo Production* case. (The House of Lords noted that the trial judge had found the motives of the employee who started the fire to be "mysteries which it was impossible to solve.") The trial judge held that the defendants were entitled to rely on the exemption clause. The Court of Appeal reversed the trial judge's decision and applied the doctrine of fundamental breach, as in *Harbutt's Plasticine*. The House of Lords expressly reversed the decision of the Court of Appeal. The House of Lords overruled the application of the fundamental breach doctrine and enforced the provisions of the exclusion clause. In an interesting discussion relating to express and implied contractual terms, the House of Lords stated:

> A basic principle of the common law of contract, to which there are no exceptions that are relevant in the instant case, is that parties to a contract are free to determine for themselves what primary obligations they will accept. They may state these in express words in the contract itself and, where they do, the statement is determinative; but in practice a commercial contract never states all the primary obligations of the parties in full, many are left to be incorporated by implication of law from the legal nature of the contract into which the parties are entering. But if the parties wish to reject or modify primary obligations which would otherwise be so incorporated, they are fully at liberty to do so by express words....
>
> Applying these principles to the instant case, in the absence of the exclusion clause which Lord Wilberforce has cited, a primary obligation of Securicor under the contract, which would be implied by law, would be an absolute obligation to procure that the visits by the night patrol to the factory were conducted by natural persons who would exercise reasonable skill and care for the safety of the factory. That primary obligation is modified by the exclusion clause. Securicor's obligation to do this is not to be absolute, but is limited to exercising due diligence in their capacity as employers of the natural persons by whom the visits are conducted, to procure that those persons shall exercise reasonable skill and care for the safety of the factory.

The doctrine of fundamental breach has been applied in Canada. For example, in 1979, the Ontario High Court of Justice decided *Murray* v. *Sperry Rand Corporation et al.*[3] The vendor of a farm

---

[3] 23 O.R. (2d) 457

harvester was prevented from relying upon a disclaimer clause in his contract with the purchaser. The "disastrous failure of the machine" to achieve the promised level of performance precluded reliance on the disclaimer. The Ontario High Court stated that the disclaimer clause in the contract would not protect the vendor against breach of a fundamental term of the contract. The court referred to a number of cases in Canadian courts where the fundamental breach doctrine has been applied, in rendering exemption clauses ineffective. The court referred specifically to an excerpt from the judgment of Lord Denning, of the Court of Appeal in England. Lord Denning also gave judgment in *Harbutt's Plasticine*.

> ... it is now settled that exempting clauses of this kind, no matter how widely they are expressed, only avail the party when he is carrying out his contract in its essential respects. He is not allowed to use them as a cover for misconduct or indifference or to enable him to turn a blind eye to his obligations. They do not avail him when he is guilty of a breach which goes to the root of the contract.

The Supreme Court of Canada referred to the overruling decision in the *Photo Production* decision in its 1980 decision in *Beaufort Realties (1964) Inc. and Belcourt Construction (Ottawa) Limited and Chomedey Aluminum Co. Ltd.*[4] The case involved the effect of a waiver of lien clause signed by the subcontractor, Chomedey Aluminum. The Court of Appeal of Ontario had found that the clause was, in effect, an exclusionary or exemption clause. The contractor had failed to pay the subcontractor and was therefore in fundamental breach of the subcontract. The issue was whether or not the exclusionary waiver clause applied in the circumstances of the fundamental breach. The clause provided:

> "ARTICLE 6. The Subcontractor hereby waives, releases and renounces all privileges or rights of privilege, and all lien or rights of lien now existing or that may hereinafter exist for work done or materials furnished under this Contract, upon the premises and upon the land on which the same is situated, and upon any money or monies due or to become due from any person or persons to Contractor, and agrees to furnish a good and sufficient waiver of the privilege and lien

4   [1980] 2 S.C.R. 718

on said building, lands and monies from every person or corporation furnishing labour or material under the subcontractor."

The Supreme Court of Canada, in referring to the *Photo Production* decision, indicated that the question of whether such a clause was applicable where there was a fundamental breach was to be determined according to the true construction or meaning of the contract (not on the *Harbutt's Plasticine* rationale that, as a rule of law, fundamental breach precluded reliance on an exemption clause). In its determination of what that true meaning was, the Supreme Court decided that the clause should not apply in the circumstances of the contractor's fundamental breach. The Ontario Court of Appeal had pointed out, in effect, that it would not be fair and reasonable for the subcontractor to continue to be bound by its waiver of lien rights as the contractor had deliberately refused to perform its basic payment obligations under the subcontract. The Supreme Court of Canada did not expressly refer to the "fair and reasonable" rationale of the Ontario Court of Appeal. Nevertheless, in the end result, the equitable outcome as determined by the Ontario Court of Appeal was preserved in the *Beaufort Realties* decision by the Supreme Court of Canada.

However, in the *Beaufort Realties* decision, even though it purported to apply the "true construction" test the court did hold that the arguably very clear waiver clause did not apply on its true construction. Therefore the contractor was able to enforce its lien. Given the somewhat confusing outcome (at least on "first reading") and the different approaches of the courts on the issue, further guidance from the Supreme Court of Canada was needed to better clarify the approach that would be taken to the enforceability of clauses that limit liability.

The more recent decision of the Supreme Court of Canada in 1989, in the case of *Hunter Engineering Company Inc. v. Syncrude Canada Ltd.*,[5] has provided that guidance. As a result of the *Hunter Engineering* decision, it has become possible to be more confident that, in certain situations, an exculpatory clause will be upheld. One clear direction to emerge from the *Hunter Engineering* decision was a unanimous acceptance of the freedom of contract and true "construction approach" over the fundamental breach doctrine, which, although overruled in the U.K., has never been

---

[5] [1989] 1 S.C.R. 426

officially overruled in Canada. Members of the Supreme Court of Canada who decided *Hunter Engineering* differed with respect to policy issues arising from the enforcement of exculpatory clauses. However, subsequent cases have further clarified the issue.

In *Hunter Engineering*, Syncrude had contracted with Hunter, and later with Allis Chalmers, for the supply of gear boxes to drive the bucket wheel conveyor belts that transported sand to Syncrude's oil extraction tar sands plant at Fort McMurray, Alberta. The contracts stipulated that Ontario law was to apply. Shortly after the gear boxes were put into service, bull gears inside them failed. Syncrude had the gear boxes rebuilt and sued Hunter and Allis Chalmers for the cost. The express warranties in the contracts had expired and were thus of no use to Syncrude. However, the implied warranty of fitness as contained in the Ontario Sale of Goods Act applied and had been breached by both Hunter and Allis Chalmers. Hunter was held liable on this basis. However, the warranty clause in the Allis Chalmers contract denied the application of all other warranties, including statutory warranties. Therefore, the only way Allis Chalmers could be held liable was if, by reason of the doctrine of fundamental breach, the exemption clause did not apply. All the members of the Supreme Court concluded that Allis Chalmers was not liable, but they took different approaches to the treatment of exemption clauses.

Chief Justice Dickson, with Justice LaForest concurring, confirmed the rule of *contra proferentem* when he held that in cases of ambiguity an exemption clause is to be strictly construed against the party relying on it. However, an exemption clause is to be given full force and effect if the language in which it is drafted is sufficiently clear to leave no doubt as to its meaning. Chief Justice Dickson did draw the boundaries of freedom of contract as follows:

> It is preferable to interpret the terms of the contract, in an attempt to determine exactly what the parties agreed. If on its true construction the contract excludes liability for the kind of breach that occurred, the party in breach will generally be saved from liability. Only where the contract is unconscionable, as might arise from situations of unequal bargaining power between the parties, should the courts interfere with agreements the parties have freely concluded.

Accordingly, Chief Justice Dickson and Justice LaForest preferred to respect freedom of contract and only consider not honouring an exculpatory clause when there is evidence of unconscionability, specifically, inequality of bargaining power, in the formation of the contract (a most unlikely possibility as far as commercial business parties are concerned).

Justice Wilson, with Justice L'Heureux-Dubé concurring, gave a larger role to the court in overturning an exemption clause in the event of a fundamental breach of contract. In their view, even if the court has decided that, at the time the contract was made, the parties succeeded in excluding liability, the court should still "import some 'reasonableness' requirement into the law" so that it "could refuse to enforce exclusion clauses in strict accordance with their terms if to do so would be unfair and unreasonable." Such reasonableness is not to be evaluated at the time the contract was entered into since the court could not possibly be privy to all the factors that influenced the course of the negotiation. Instead, the courts should "determine *after the particular breach has occurred* whether an exclusion clause should be enforced or not." According to Justice Wilson, the courts should decide "*whether, in the context of the particular breach which had occurred* it was fair and reasonable to enforce the clause in favour of the party who had committed the breach even if the exclusion clause was clear and unambiguous."

Justice Wilson did not say that every fundamental breach will result in an exemption clause losing its validity, rather the exemption clause would only lose its validity if enforcing it is not fair and reasonable.

In the years since the *Hunter Engineering* decision the courts have generally adopted the approach advocated by Chief Justice Dickson over that of Justice Wilson. Clear and direct exemption clauses found in contracts negotiated between commercial parties of relatively equal bargaining power have, virtually without exception, been upheld. Accordingly, it is now possible to say with far more certainty than had previously been the case, that a properly drafted exculpatory clause may be relied on. The Canadian court emphasis is now to look carefully at the wording of each contract, even in circumstances involving a fundamental breach, and resolve matters according to the true intention of the parties at the time the contract was negotiated.

To generalize further on the status of the doctrine of fundamental breach in Canada, as succinctly summarized by one of the writer's colleagues after researching the issue: "the doctrine of fundamental breach is not yet officially dead in Canada but it appears to be mortally wounded," at least as far as engineering project cases are concerned.

Hence, an exemption clause will be scrutinized by a strict construction of its wording. If the clause is clear and direct it will be enforced unless it is unconscionable. The courts have been very unwilling to find that agreements between commercial parties are unconscionable. Clearly drafted exculpatory clauses should, generally, be effective in contracts involving engineering projects and services.

As a practical matter, the engineer in business should pursue the negotiation of exemption clauses limiting liability where appropriate, in terms as advantageous as possible.

SAMPLE CASE STUDY

The following hypothetical case study and answer is included for illustrative study purposes.

1. Prime Equipment Inc. ("Prime") is a company engaged in the business of supplying heavy equipment used in the oil exploration and drilling industry.

Prime had become aware that Conventional Oil Company Ltd. ("Conventional") required a contractor to design, manufacture, supply, and install specialized gear boxes. The gear boxes would be used to drive a number of bucket wheel conveyor belts that transported sand at Conventional's oil extraction tar sands plant in Alberta. Prime decided to tender on the Conventional contract.

In order to tender on the contract, Prime set out to purchase the gear boxes. Prime was contacted by a representative of MachineWorks Ltd., a company which manufactured similar equipment. After visiting Conventional's site and examining the conveyors, the representative of MachineWorks became familiar with the requirements of the gear boxes. MachineWorks represented to Prime that MachineWorks would be able to design and manufacture the specialized gear boxes and that the gear boxes would be suitable for the purpose intended. On the basis of these representations, MachineWorks and Prime entered into a contract. MachineWorks agreed that if Prime were successful in its tender to Conventional, MachineWorks would provide the equipment for a price of $300,000. The contract also contained a provision limiting MachineWorks' total liability to $300,000 for any loss, damage, or injury resulting from MachineWorks' performance of its services under the contract.

Based on the information provided by the MachineWorks representative, Prime prepared and submitted its tender to Conventional. Conventional accepted the tender and entered into a contract with Prime for the gear boxes.

The gear boxes were installed at Conventional's site by employees of Prime according to MachineWorks' installation procedures. Shortly after the gear boxes were put into service, main gears inside them failed. As a result of this failure, the conveyors were damaged and it was impossible for Conventional to operate its conveyors. MachineWorks made several unsuccessful attempts to correct the gear boxes.

In order to meet its obligations under the Conventional contract, Prime hired another supplier to correct the defects in the gear boxes. For an additional $400,000 Prime was able to correct the problem by replacing the gear boxes with gear boxes manufactured by another company and by repairing the damage to the conveyors. The total amount that had been paid by Prime to MachineWorks prior to discovering the defects was $250,000.

Explain and discuss what claim Prime can make against MachineWorks in the circumstances. Would Prime be successful in its claim? Why? In answering, please include a summary of the development of relevant case precedents.

Sample Answer:

The facts reveal that there was a major or fundamental breach of a contract. The contract included an exemption clause, that is a clause limiting liability "to $300,000 for any loss, damage, or injury resulting from MachineWorks performance of its services under the contact." These factors point to the issue of whether or not the exemption clause is enforceable and the amount of the claim that can be made against Prime in the circumstances.

At one time, Canadian courts followed English precedent in applying what was called the "fundamental breach doctrine," a concept that rendered an exemption clause unenforceable in the circumstances of a fundamental breach, that is a breach of such significance as to go directly to the root of the contract itself (as pointed out in the *Harbutt's Plasticine* case).

Subsequently, the doctrine of fundamental breach was officially overruled in England (the *Photo Production* case). Although never officially overruled in Canada, the Supreme Court of Canada has now clarified in the *Hunter* v. *Syncrude* case, its very strong preference not to follow the doctrine of fundamental breach in commercial cases such as this, but rather to follow the reasoning of the U.K. courts that overruled the doctrine of fundamental breach in England, in applying the "true construction approach." Accordingly, Canadian courts no longer automatically disregard a clause limiting liability in the circumstances of a fundamental breach of contract. Instead, pursuing the principle that business parties are free to choose whatever commercial terms they agree upon, in the circumstances of any breach of contract be it fundamental or otherwise, the courts will look at the wording of the contract and make a determination on whether or not the parties at the time they had executed the agreement had intended that the limitation of liability provision in the contract would be applied in the circumstances of the breach that had ultimately occurred.

(Note: reference to *Harbutt's Plasticine, Photo Production*, and *Syncrude and Hunter* embellish the answer, but are not essential).

In calculating damages for breach of contract, the amount is determined as the amount arising naturally from the breach or as may reasonably have been contemplated by the parties at the time they entered into the agreement. Here the agreement evidenced that the original contract price for the equipment was $300,000 and that MachineWorks' total liability for any loss, damage, or injury was also limited to $300,000. Ultimately, Prime paid a total of $650,000 when it ought to have received what it contracted for, for $300,000. Accordingly, its damages amounted to the excess, being $350,000.

Given the decision of the Supreme Court of Canada in *Hunter and Syncrude*, the clear wording of the provision limiting liability would stand, even in the circumstances of the fundamental breach, and Prime would only be able to recover $300,000 from MachineWorks of the $350,000 in damages it sustained as a result of MachineWorks' breach of contract.

CHAPTER TWENTY-ONE

THE AGREEMENT BETWEEN CLIENT AND ENGINEER

A contract between a client and a professional engineer must include all the essential contract elements.

A contract between a client and an engineer will not usually specify the measure of the standard of care in performance that is expected of the engineer. The contract will simply state that the engineer is to provide engineering services in connection with a particular project. The document may detail the scope of such services, but it will not necessarily specify the degree of care that is required of the engineer in carrying out those services. That degree of care will be an implied term in the contract. As pointed out earlier, an engineer is liable for incompetence, carelessness, or negligence that results in damages to the client; the engineer is responsible as a professional for not performing with an ordinary and reasonable degree of care and skill. The standard of performance expected of the engineer in contract is essentially the same as the standard expected in tort law, unless otherwise provided by a particular contract.

Most engineers are likely aware of the frequency of court actions against professionals. The potential magnitude of damage claims flowing from breach of contract for negligently performed engineering services can be enormous! Professional engineers should carry appropriate and adequate professional liability insurance coverage.

THE AGENCY RELATIONSHIP

By entering into a client-engineer agreement, the engineer usually enters into an agency relationship in which the client is the principal and the engineer is the agent. As an agent, the engineer must be careful to act only within the scope of his or her authority as agreed upon with the principal. If an engineer exceeds the scope of his or her authority, the engineer may be liable to the principal for damages resulting from the engineer's actions. For example, suppose an engineer acts as agent on behalf of a client to negotiate the terms of a construction contract. The engineer should seek the client's approval of all terms before confirming acceptance of the contract, unless the client has otherwise clearly authorized the engineer.

THE ENGINEER'S REMUNERATION

The client-engineer agreement should be in written form. The contract should clearly outline the nature of the project, the nature of the services that are required, the basis of payment, and all general terms and conditions mutually acceptable to both client and engineer. If an engineer is retained by a client to perform services, and undertakes to do so without agreeing with the client on the amount of remuneration to be received, the law implies that the engineer shall be paid a reasonable amount for services on a *quantum meruit* basis. A suggested schedule of fees, published by a provincial engineering association, should provide a reasonable basis on which to determine remuneration.

ESTIMATED FEE

When an engineer enters into a contract with a client, he or she should cautiously estimate, if required and appropriate, the amount of the engineer's total fee. The engineer should also emphasize to the client that the quoted total fee is only an estimate. The issue of discrepancies between estimates and final costs has come before the courts. An example is the 1979 decision of the Ontario High Court of Justice in *Kidd* v. *Mississauga Hydro-Electric Commission et al.*[1] In proposing to undertake a study, a consulting engineer estimated that the cost of his support staff would be $5,000.00. The actual cost of the support staff was $14,447.00. The consulting engineer brought an action to recover the full cost of the support staff. The court held that, in the circumstances, the engineer's client was entitled to rely upon the engineer's exper-

---

[1]   97 D.L.R. (3d) 535

tise in making the estimate. The following excerpt from the court's judgment is of interest:

> ... I consider that the reference to $5,000.00 has a major effect upon the liability of the defendant. Admittedly the plaintiff could not give a precise estimate of the cost of support staff, but he was an experienced consultant — and even if he were not, it surely is incumbent upon a professional man to estimate the cost of services at something closer to the eventual figure than was done here.... I do not, of course, mean to say that all estimates are necessarily binding. Clearly they are not, and the plaintiff here might well have been allowed, because of the vagueness of his estimate, a substantial margin of error. But where the eventual figure is almost three times the original estimate, it is my view that the estimator should be held to that original figure.
>
> I can find no authority directly supporting the conclusion just reached. Normally if the contract fails to set a price (and certainly no firm price was set here) the Courts will set a reasonable price upon a *quantum meruit* and it is here conceded that, the merit of the study aside, the support staff work was done and the charges therefore were appropriate. But the essential distinction here is that a figure was mentioned, that it was wildly inadequate, and the defendant committed itself to payment because that figure was mentioned. This may be no more than another example of the negligent misrepresentation principle enunciated in *Hedley Byrne & Co. Ltd.* v. *Heller & Partners Ltd.*, [1964] A.C. 465. The duty upon the representor was there expressed by Lord Reid at p. 486, as follows:

> A reasonable man, knowing that he was being trusted or that his skill and judgment were being relied on, would, I think, have three courses open to him. He could keep silent or decline to give the information or advice sought; or he could give an answer with a clear qualification that he accepted no responsibility for it or that it was given without that reflection or inquiry which a careful answer would require; or he could simply answer without any such qualification. If he chooses to adopt the last course he must, I think, be held to have accepted some responsibility with the inquirer which requires him to exercise such care as the circumstances require.

> In the case at bar it seems to me the plaintiff adopted the third course, In the light of (a) the plaintiff's familiarity with reports

of this nature and the work involved, and (b) the eventual cost of the support staff work, I cannot conceive that the answer was given carefully. This is an action in contract for payment not one in tort for negligence, but the obligations of the plaintiff and the consequences for breach must be at least as great.

It may be also that the contract can be viewed as one for the payment of support staff at a cost of approximately $5,000, and because the charge was not close to that figure the Court will give judgment for the only figure mentioned....

STANDARD-FORM ENGINEERING AGREEMENTS

Recommended forms of agreement for professional engineering services between client and engineer and between engineers and other consultants are available from provincial associations, such as the Association of Professional Engineers of Ontario, and from other sources, such as the Association of Consulting Engineers of Canada. The standard-form engineering agreements currently available have been carefully developed. The forms set out very good basic contract formats. Some require more detail of completion than others with respect to project definition, the description of engineering services to be provided, and the engineer's fee for services rendered. The general terms and conditions set out in the forms cover important aspects of the contractual relationship between the parties. But additional terms, conditions, or modifications to the standard form may be necessary. A standard-form contract may constitute a satisfactory form of agreement between parties, but the standard-form contract may not always suffice. Each contract should be "tailored" so that it accurately records the particular agreement between the parties, a drafting process that is usually best left to lawyers. The importance of ensuring that contracts are competently drafted cannot be overemphasized.

LIMITING LIABILITY BY CONTRACT

The doctrine of fundamental breach has been discussed, and we have considered its applicability in Canadian cases. Any contract must take the doctrine into account. There is no reason a contract between an engineer and client cannot contain a provision whereby the engineer limits, to some extent, his or her liability for damages resulting from performance of engineering services. Such a provision may well be totally unacceptable to the client, however. Insistence upon an unreasonable provision may well result in the loss of a prospective client. An example of an unreasonable clause might be one that limits liability to an amount less than the engineer's fee for the project.

A more reasonable approach is to limit the engineer's potential liability to the extent of his or her professional liability insurance coverage. If such an approach is taken, the details of the liability coverage should be communicated to the client, so that the client can assess the appropriateness of the coverage.

ENGINEER'S COMPLIANCE WITH THE LAW

An engineer who provides engineering services is expected to comply with common-law principles relating to tort and contract. In addition, he or she must comply with all statutes and regulations applicable to the particular nature of engineering services performed. The engineer is expected to have a reasonable knowledge of applicable statutes and regulations. He or she should ensure an awareness of requirements by investigating applicable federal and provincial laws relevant to the industry within which particular engineering services are being provided.

There are many federal and Ontario statutes of a technical nature. The engineer should maintain up-to-date knowledge of relevant statutes and regulations. Government sources and provincial engineering associations should be of assistance. Sometimes legal advice on particular matters may be required. The following is a list of technical statutes that provide for engineering involvement in some respects or that are relevant to engineering practice. The list is illustrative rather than exhaustive:

1. Aggregate Resources Act, R.S.O. 1990, c. A. 8

2. Amusement Devices Act, R.S.O. 1990, c. A. 20

3. The Boilers and Pressure Vessels Act, R.S.O. 1990, c. B. 9

4. The Building Code Act, R.S.O. 1990, c. B. 13

5. Charitable Institutions Act, R.S.O. 1990, c. C. 9

6. Child and Family Services Act, R.S.O. 1990, c. C. 11

7. Community Recreation Centres Act, R.S.O. 1990, c. C. 22

8. The Construction Lien Act, R.S.O. 1990, c. C. 30

9. Consumer Packaging and Labelling Act (Canada), R.S.C. 1985, c. C-38

10. Day Nurseries Act, R.S.O. 1990, c. D. 2

11. Developmental Services Act, R.S.O. 1990, c. D. 11

12. The Drainage Act, R.S.O. 1990, c. D. 17

13. Elderly Persons Centres Act, R.S.O. 1990, c. E. 4

14. Elevating Devices Act, R.S.O. 1990, c. E. 8

15. Energy Act, R.S.O. 1990, c. E. 16

16. The Environmental Assessment Act, R.S.O. 1990, c. E. 18

17. The Environmental Protection Act, R.S.O. 1990, c. E. 19

18. Fire Marshals Act, R.S.O. 1990, c. F. 17

19. The Hazardous Products Act (Canada), R.S.C. 1985, c. H-3

20. Homes for the Aged and Rest Homes Act, R.S.O. 1990, c. H. 13

21. Homes for Retarded Persons Act, R.S.O. 1990, c. H. 11

22. Lakes and Rivers Improvement Act, R.S.O. 1990, c. L. 3

23. The Mining Act, R.S.O. 1990, c. M. 14

24. The Municipal Act, R.S.O. 1990, c. M. 45

25. The Occupational Health and Safety Act, R.S.O.1990, c. O. 1

26. The Occupiers' Liability Act, R.S.O. 1990, c. O. 2

27. The Ontario New Home Warranties Plan Act, R.S.O.1990, c. O. 31

28. Ontario Water Resources Act, R.S.O. 1990, c. O. 40

29. The Operating Engineers Act, R.S.O. 1990, c. O. 42

30. Petroleum Resources Act, R.S.O. 1990, c. P. 12

31. The Planning Act, R.S.O. 1990, c. P. 13

32. Professional Engineers Act, R.S.O. 1990, c. P. 28

33. Public Transportation and Highway Improvement Act, R.S.O. 1990, c. P. 50

34. Vocational Rehabilitation Services Act, R.S.O. 1990, c. V. 5

An examination of the list will indicate the diversity of statutes with which an engineer may be required to be familiar. The statutes to be complied with will vary according to the nature of the contract. For example, in a contract dealing with real property (land), requisite approvals may be required to subdivide land; and relevant statutes may include the Planning Act and the Municipal Act. (Legal advice should be obtained where interests in real property are involved.)

In addition to statutes, there are many codes and building regulations with which the engineer must comply. For example, electrical codes, provincial building codes, and municipal by-laws that relate to structural, fire, health, and other safety requirements. Compliance with municipal zoning or licensing by-laws may also be required.

In fulfilling a contract, the engineer must be careful to comply with all applicable laws; otherwise, serious problems may arise. Failure to comply with building permit requirements, for example, may expose the engineer's client to substantial additional cost or other loss, for which the engineer may be responsible.

As already indicated, an engineer should be especially cautious when involved in a contract dealing with real property. The body of law concerned with land is singularly complex. Much of it is derived from concepts developed in feudal England. As a result, the laws tend to be filled with archaic terms and somewhat unusual concepts. It is extremely important to obtain legal advice where interests in real property are involved.

Although a comprehensive review of the law of real property is outside the scope of this text, there are some aspects of it that should be of particular interest to engineers. For example, many governments have actively attempted to plan the growth of communities they oversee. In Ontario, this desire for planned growth has led to a system of "subdivision control." The Planning Act of Ontario provides that, with certain exceptions, land shall not be divided (that is, sold in separate parts to different parties) unless a "consent" under the Planning Act has been obtained or the sale is in accordance with a "registered plan of subdivision." When the owner wants to divide land into parcels, he or she seeks a consent to the "severance" from the appropriate government body. This duty is often performed by the "committee of adjustments" for the municipality where the land is located.

When a person (for example, a land developer) buys one large parcel of land and wants to sell it in many small parts, the owner usually registers a plan of subdivision. The owner submits a detailed plan to the municipality. Generally, this is a lengthy process and involves negotiations between the developer and the municipality; also a variety of government agencies usually consider the plan.

The Planning Act is not limited to commercial land developers; rather, it applies whenever someone seeks to sell a piece of land that "abuts" (touches) other land owned by the same person.

The penalties for violating the Planning Act are severe. Section 49(21) of the Planning Act provides that any sale or agreement that contravenes the Act fails to "create or convey any interest in land." The effect of this section is dramatic. Suppose A "buys" land from B, who owns a piece of abutting land. If A and B have not obtained a consent, and if the sale is not in accordance with a plan of subdivision, A and B have violated the Planning Act. Section 49(21) of the Act effectively "voids" the sale, so A does not own the land, even if he paid for it. Recovering the purchase price from B may prove difficult or impossible. To avoid this problem, A's lawyer should find out who owns adjoining land when he or she performs the "search."

The control of land subdivision is one of the methods used by government to control development. However, the Planning Act allows municipalities to review development even if no division of land is involved. A municipality may designate an area as one where "site-plan control" is in effect. In a site-plan-control area, detailed plans for any development must be approved by the municipality. Often municipalities demand that plans for particular projects be modified.

Following is a list of provincial statutes relating to the control of developments:

Alberta: Planning Act, R.S.A. 1980, c. P-9

British Columbia: Municipal Act, R.S.B.C. 1979, c. 290

Manitoba: Planning Act, R.S.M. 1987, c. P. 80

New Brunswick: Community Planning Act, R.S.N.B. 1973, c. C-12

Newfoundland: Urban and Rural Planning Act, R.S.N. 1990, c. V-7

Nova Scotia: Planning Act, R.S.N.S. 1989, c. 346

Prince Edward Island: Planning Act, R.S.P.E.I.1988, c. P-6

Quebec: Land Use Planning and Development Act, R.S.Q. c. A-19.1

Saskatchewan: Planning and Development Act, R.S.S. 1978, c. P-13.

CHAPTER TWENTY-TWO

CONCURRENT LIABILITY IN TORT AND CONTRACT

Unless otherwise stated in a contract, the standard of care expected of an engineer in performing services pursuant to a contract is the same standard of care by which an engineer's performance is measured in tort. This "overlap" raises the question of whether an engineer can be concurrently liable for both breach of contract and tort. The consequences of simultaneous liability raise some rather startling possibilities. For example, the measure of damages for breach of contract may not always be the same as the measure of damages arising in tort. And limitation periods may differ in contract and in tort.

An illustration is provided by the 1967 decision in *Schwebel v. Telekes.*[1] The action against a notary public alleged negligence in acting for the plaintiff in the purchase of a home. The Ontario Court of Appeal concluded that the plaintiff's action was not in tort but in contract; the court noted that the duty of care arose by virtue of a contractual relationship, and had no existence apart from that relationship. The court concluded that, where breach of contractual duty is involved, the limitation period of six years runs from the breach of duty, rather than from the time that breach was or ought to have been discovered. In tort, however, as the Ontario Court of Appeal noted in 1976 in *Dominion Chain Co. Ltd.* v. *Eastern Construction Co. Ltd. et al.,*[2] and as the Supreme Court of Canada confirmed in 1984 in *City of Kamloops* v. *Nielsen et al.,*[3] the limitation period starts to run not when the services are performed, but when the damage is first detected or ought to have been detected.

[1]   (1967), 10 O.R. 541
[2]   [1974] 3 O.R. (2d) 481
[3]   [1984] 29 C.C.L.T. 97

Questions may also arise with respect to the respective liabilities of concurrent tortfeasors. Suppose, for example, that one of the tortfeasors has precluded liability by virtue of a contract with the plaintiff. The point is illustrated in the *Dominion Chain* Case. Dominion Chain Co. Ltd. entered into separate contracts with a contractor and with an engineer in connection with the construction of a factory. The factory roof developed very serious leaks five years after construction. Dominion Chain initiated an action against the contractor and the engineer, but the trial judge dismissed the action against the contractor because a clause in the contract limited the contractor's liability. As owner, Dominion Chain had, in the contract, waived its right to make claims against the contractor after the guarantee period had expired. Although the contractor was not contractually liable to the owner because the guarantee period had expired, the trial judge concluded that the damage to the roof was, however, the result of negligence by both the contractor and the engineer, and apportioned 75 percent and 25 percent respectively. The trial judge then applied Section 2 of The Negligence Act of Ontario.[4] The Act provides that where two persons are found at fault or negligent, they are jointly and severally liable to the person suffering the damage. Between themselves, however, in the absence of any contract, each is liable to make contribution and indemnify the other in the degree in which they are respectively found to be at fault or negligent. The trial judge concluded, therefore, that as only the engineer was required to satisfy the owner's damages, the contractor should contribute to the engineer 75 percent of such damages. The trial judge's decision was appealed. The Ontario Court of Appeal[5] did not agree with the trial judge. Noting that a contractor and an engineer may be liable in tort as well as in contract for negligent performance of contractual duties, the Court of Appeal held that the contractor was not required to contribute 75 percent of the damages to the engineer. The court pointed out that Section 2 of The Negligence Act did not apply unless both the contractor and the engineer, as joint tortfeasors, were liable to the plaintiff, Dominion Chain Co. Ltd. The contractor had escaped such liability because of its contract with the plaintiff, thus it was not liable to contribute 75 percent of the damages to the engineer.

---

[4]   R.S.O. 1980, c. 315
[5]   68 D.L.R. (3d) 385

The decision of the Court of Appeal in *Dominion Chain* was appealed by the engineer. The Supreme Court of Canada[6] concluded that, because of the contract between the owner and the contractor, the contractor was not liable to make contribution to the engineer. However, the court focussed on contractual provisions to decide the appeal; there was no direct consideration of the question of concurrent liability in tort and contract.

Hence, in *Dominion Chain*, the Ontario Court of Appeal concluded that an engineer or a contractor may be concurrently liable in tort as well as in contract. But the Supreme Court of Canada has not made a decision on the principle of concurrent liability. It did so subsequently, however, in its 1986 judgment in *Central Trust Co.* v. *Rafuse,*[7] a case involving solicitors. The principal issue considered by the Supreme Court of Canada was whether or not a lawyer could be liable to a client in tort as well as in contract for damages for failing to meet the appropriate standard of care for which the lawyer had been retained. The Supreme Court concluded that the solicitor could be liable in tort as well as in contract. The tort liability then exposed the solicitor to the tort limitation period of six years from "discoverability."

Design deficiencies relating to an overhead contact system in a tunnel for electrically powered locomotives resulted in another tort and contract case in 1988. The case was *B.C. Rail Ltd.* v. *Canadian Pacific Consulting Services Ltd. et al.* (1988) 29 C.L.R. 30. The contract expressly provided that the services were to be performed with reasonable skill, care, and diligence. The contractor was bound to ensure that its subcontractors performed to the same standards. The design of this overhead contact system was subcontracted. The subcontractor carried out no testing or data gathering of its own inside the tunnel and did not request copies of underlying reports. It was not informed of a large volume of water percolating through the tunnel rock or the presence of sulphur compounds.

Within 14 months extensive damage to the overhead contact system in the tunnels was discovered. The water seepage produced a very humid atmosphere that promoted stress corrosion cracking damage. Hydrogen sulphide, ammonia, and nitrites contributed to promote cracking across the wire. The owner brought an action for damages for breach of contract and negligence. The

---

[6] [1978] 2 S.C.R. 1346
[7] [1986] 2 S.C.R. 147

failure to gather the necessary data was regarded as a clear breach of the duty of care imposed by the contract. The owner was entitled to recover the reasonable costs of redesigning and rewiring the system as well as the cost of future replacement of the wire.

Another important decision from the Supreme Court of Canada, involving concurrent liability in tort and contract was its 1993 decision in *B.G. Checo International Ltd.* v. *British Columbia Hydro and Power Authority*.[8] The case involved a contract to install transmission lines. It was a term of the contract that was expressly spelled out that the right of way had been cleared. This was a negligent misrepresentation by the utility in the tender documents. The right of way had not been adequately cleared. There was still a fair amount of debris in the right of way. It was substantiated that B.C. Hydro knew before the tenders closed that the contractor employed to clear the right of way hadn't done the job properly. The court held that B.C. Hydro had a duty to tell the tenderers that the right of way had not been adequately cleared and it was negligent for not doing so. B.C. Hydro was sued and found concurrently liable in contract and in tort.

Accordingly, any previous uncertainty about the application of the concept of concurrent liability in tort and contract in Canada has been put to rest.

---

[8]   [1993] 2 W.W.R. 321

CHAPTER TWENTY-THREE

THE DUTY OF HONESTY

When an engineer enters into a contract, he or she assumes a duty of care in performing services. Implicit in that duty of care is the duty of the engineer to act with absolute honesty.

The penalties for dishonesty can be severe. Where fraud is involved, a contract may be repudiated and damages may be awarded for the tort of deceit. In addition, fraud is a criminal offence; it is punishable, on conviction, by imprisonment for up to ten years. Section 380 of the Criminal Code[1] (Canada) provides, in part:

> (1) Every one who, by deceit, falsehood or other fraudulent means, whether or not it is a false pretence within the meaning of this Act, defrauds the public or any person, whether ascertained or not, of any property, money or valuable security, (a) is guilty of an indictable offence and is liable to a term of imprisonment not exceeding ten years....

The engineer is usually retained as the agent of the client. A relationship of trust exists between an agent and his or her principal; the duty of good faith that arises from that trust is a very important one. Section 426 of the Criminal Code deals with violation of the principal–agent relationship. The section deals with secret commissions — bribes and kickbacks. It provides:

> (1). Every one commits an offence who
>
> (a) Corruptly
>
> (i)   gives, offers or agrees to give or offer to an agent, or

---

[1] R.S.C. 1985, c. C-46

(ii) being an agent, demands, accepts or offers or agrees to accept from any person, any reward, advantage or benefit of any kind as consideration for doing or forbearing to do, or for having done or forborne to do, any act relating to the affairs or business of his principal or for showing or forbearing to show favour or disfavour to any person with relation to the affairs or business of his principal; or

(b) with intent to deceive a principal, gives to an agent of that principal, or, being an agent, uses with intent to deceive his principal, a receipt, account or other writing

(i) in which the principal has an interest,

(ii) that contains any statement that is false or erroneous or defective in any material particular, and

(iii) that is intended to mislead the principal.

A conviction for taking secret commissions can result in imprisonment.

Various sections of the Criminal Code and sections of other statutes — such as the Income Tax Act (Canada) and the Competition Act (Canada) — provide substantial sanctions to deal with dishonesty. The sanctions include fines and imprisonment, and they emphasize the importance of honesty and integrity in Canadian business dealings.

The engineer should also be aware that, under section 121 of the Criminal Code, it is an offence for a government employee to accept any gift from a person who has dealings with that government, unless the head of the employee's government branch consents in writing. It is also an offence to give any such gift or confer a benefit on a government employee if the person giving the gift or conferring the benefit has dealings of any kind with the government. The offence is punishable by imprisonment for up to five years.

CHAPTER TWENTY-FOUR

CONSTRUCTION CONTRACTS

The engineer is not normally a party to a construction contract; the engineer usually has a separate contract with the owner (client). Under the separate contract, the engineer may undertake to prepare plans and specifications, to assist in the tendering process, and to administer the construction contract between the owner and the contractor.

As administrator, the engineer may make decisions of major significance to the rights and obligations of the owner and the contractor. For example, many construction contracts provide that the engineer shall, in the first instance, interpret the provisions of the contract; and be the judge of the performance of the respective obligations of the parties to the construction contract. The words "in the first instance" are included for the purpose of reaching timely decisions during construction, yet leaving to a more appropriate forum — for example, a court or arbitrator — the ultimate decision as to how the contract ought to be interpreted. Some examples of aspects of administration of construction contracts with respect to which the engineer is often authorized by both the owner and the contractor to make decisions are as follows:

(1) in the preparation of payment certificates;

(2) in the preparation of certificates to evidence substantial and final completion of the work;

(3) in the event of delays during construction, the engineer may determine the appropriateness of extending time for completion;

(4) in the determination of whether the contractor has failed to fulfil his or her obligations to an extent that would justify termination of the construction contract;

(5)   in the valuation of changes or "extras" to the contract, the engineer may decide appropriate changes in contract price and in time for completion,

(6)   in determining whether actual subsurface conditions differ materially from those conditions described in the plans or specifications for the work;

(7)   in the event of emergencies, the engineer is usually authorized to direct the contractor's work methods to ensure safety to property and workers;

(8)   to inspect the progress of the construction and to reject work that does not comply with the contract documents.

The manner in which an engineer ought to discharge decision-making powers has been the subject of court decisions. In 1953, the Ontario Court of Appeal decided *Brennan Paving Co. Ltd.* v. *Oshawa.*[1] The court considered the conduct of an engineer who acts as agent for the owner, and who also acts as certifier of payment certificates. The court concluded that as certifier, the engineer is required to act judicially and in an independent and unbiased manner; the engineer should not then act as agent on behalf of the engineer's principal. The court stated, in part:

> Where, as here, the engineer's certificate is a condition precedent to payment, the engineer occupies two positions: first, one as agent of the owner under the contract; second, a quasi-judicial position as certifier between the parties: *Hudson on Building Contracts*, 7th ed., p. 286. The two positions are distinct and separate. Different duties attach to them and different consequences flow from the performance or breach of those duties. Under the law of principal and agent he may, within the scope of his duties, bind his principal. As certifier deciding between the parties he must act judicially.... To act judicially as certifier requires him, where the question arises, to consider and give effect to any conduct on his part as agent vis-à-vis the contractor which has bound the owner as his principal to the advantage of the contractor. In this connection he must act qua certifier as independently as if some other person rather than himself had been the agent of the owner under the contract. All that seems crystal clear to me.

[1]   [1953] 3 D.L.R. 16

In 1960, the Supreme Court of Canada decided *Kamlee Construction Ltd. v. Town of Oakville.*[2] The construction contract provided that the "decision of the engineer shall be final and binding upon both the contracting parties as to the interpretation of the specifications and as to the material and workmanship." The contractor disagreed with the decisions of the engineer. A "battle of wills" developed between the engineer's representatives and the contractor's representatives; the factions disagreed on many basic questions concerning method and detail. The Supreme Court of Canada held that the contractor was not entitled to repudiate the contract on the basis of the engineer's conduct. Under such a contract, the court held, an engineer is required to act "judicially"; the engineer must make decisions dictated by the engineer's own best judgment. He or she must decide the most efficient and effective way to carry out the contract; the engineer's judgment must not be affected by the fact that he or she is being paid by the owner.

The question of the manner in which the engineer carries out duties was also considered in another case: the 1959 decision in *Croft Construction Co. v. Terminal Construction Company.*[3] Contracting parties had agreed that payment to the contractor would be based on calculations by the engineer. The court upheld that the engineer's figures would govern the payments, even if the engineer had made an honest mistake. The Ontario Court of Appeal stated in part:

> It is well considered that a computation of the character in question is conclusive and binding upon the parties in the absence of fraud or bad faith or, unless the person entrusted with the duty of making it, has knowingly and wilfully disregarded his duty. It is not suggested here that there was any fraud or bad faith on the part of the engineer....

The immunity of a certifier from any loss caused by lack of skill has come into question. In 1974, the House of Lords decided *Sutcliffe v. Thackrah et al.*[4] An architect who acted as the owner's agent in preparing payment certificates was, the House of Lords held, under a duty to the owner to exercise care; the architect was, in fact, liable to the owner for negligent over-certification unless he could show that he was acting as an

[2] [1961] 26 D.L.R. (2d) 166
[3] 20 D.L.R. (2d) 247
[4] [1974] All E.R. 859

arbitrator in preparing the certificates. The architect was unable to substantiate that he was acting as an arbitrator; thus he was liable for any loss caused by his lack of skill or negligence. The House of Lords concluded that immunity would be extended only to an arbitrator. In his capacity as certifier on behalf of the owner, immunity was not available to the architect.

To summarize, where an engineer is empowered to make decisions that are final and binding upon the parties to a construction contract, the engineer must act judicially notwithstanding that he or she was originally retained by the owner. In acting judicially, the engineer must act independently of the owner (his or her principal) and in good faith. Provided this is the case, the engineer's decisions will be binding upon the parties. In addition, an engineer may be liable for loss caused by negligent certification, if future Canadian court decisions apply *Sutcliffe* v. *Thackrah.*

CERTIFICATES FRAUDULENTLY PREPARED

Canadian courts have heard cases that involve fraudulently prepared engineers' certificates. In *Grant, Smith & Co.* v. *The King,*[5] certificates prepared by an engineer in collusion with a drilling contractor represented a fraudulent attempt to overstate rock quantities. The certificates were set aside by the court.

INSPECTION SERVICES

In administering a construction contract, an engineer normally acts as an inspector. In doing so the engineer must discharge all duties competently and must demonstrate a reasonable degree of care and skill.

Unless the contract provides otherwise, the engineer will be expected to inspect all significant aspects of construction. The inspection may be done personally by the engineer, or by competent representatives selected by the engineer. The courts expect a high standard from professionals; the standard is illustrated in a 1976 decision of the Ontario Court of Appeal in *Dabous* v. *Zuliani et al.*[6] An architect entered into a contract for the design and supervision of the construction of a house. During the construction, a metal chimney was installed too close to wooden joists;

---

[5] [1920] 19 Ex. C.R. 404
[6] 68 D.L.R. (3d) 414

the oversight eventually resulted in fire damage to the house. The court considered the architect's inspection of the chimney installation, and stated:

> Prior to the installation of this prefabricated chimney which caused the fire, the builder had installed another prefabricated chimney in this house which serviced the furnace in the basement. This chimney was installed in direct contact with the wooden joist at the first-floor level and was observed by the architect who cautioned the builder against this practice and directed correction. The prefabricated fire-place was late in arriving and because of some urgency in completing the building, the fire-place together with its chimney was installed on a week-end and the area where the prefabricated chimney passed through the second floor level was boxed in and covered with gyprock. As a result, the offending installation was not observed by the architect. It should perhaps be added that the architect took no steps to have some of the gyprock removed, so that he might examine the chimney installation, although that would have involved trifling cost....
>
> The proper installation of this type of chimney was critical to the safety of the dwelling. The fact of a previous similar improper installation and its correction should have magnified the concern of the architect rather than diminished it as was argued on their behalf. The failure of the defendant architects to ensure that the chimney leading from the fire-place was properly installed, even if this required the removal of some of the concealing gyprock, was a failure to meet the reasonable standard of care which they owed to the plaintiff and the appeal of the architects against their liability was therefore dismissed....

The decision may seem harsh to some. However, it illustrates the very high degree of care expected by the courts.

THE ENGINEER'S ADVICE TO THE CONTRACTOR

In a construction contract, the engineer is normally expected to inspect rather than actually supervise the work methods of the contractor. Most construction contracts provide that the contractor shall have complete control of construction methods and procedures. However, consistent with tort law principles, the engineer may be expected to advise a contractor in circumstances where the contractor is reasonably relying on the engineer's expertise. This may well require the engineer to make a difficult judgment call; and the engineer must be extremely careful in doing so as unwarranted interference with a contractor's work methods may give rise to a damage claim.

Some contracts provide that the engineer is to approve the contractor's work methods, in which case the engineer's obligations are more clear. To illustrate, in 1977 the Supreme Court of Canada decided *Demers et al.* v. *Dufresne Engineering et al.*[7] During the construction of bridge piers, the contractor failed to use sufficient vertical reinforcing steel in constructing a caisson. The caisson exploded under pressure and had to be rebuilt at a cost of $1,400,000. The Supreme Court of Canada held the engineer 50 percent liable for the damage. In part, the court stated:

> The explosion of the caisson was due to a glaring error in the method of performing the work that was selected by the contractor; having failed to take the low resistance of concrete in tension into account, the latter did not provide for the use of vertical reinforcing steel. The engineer was aware of this incorrect method of doing the work; if he had not been aware of it, I would have had no hesitation in saying that he ought to have been since this was such an enormous error. By remaining silent, the engineer implicitly approved the work method chosen by the contractor. Moreover, he also implicitly approved the minor alteration which consisted in adding a small quantity of vertical reinforcing steel and which, even having regard to the preliminary calculations made by his representative ... was obviously inadequate. By committing these two errors the engineer effectively allowed the work to be performed incorrectly, and this caused the accident. The contractor's error indicates how much he was in need of the engineer's guidance in order to perform the work properly; this need for guidance gave rise to the engineer's obligation to give it to the contractor, to see, in short, that the error be corrected. By failing to carry out this contractual obligation, the engineer became liable toward the contractor.

[7]   [1979] 1 S.C.R. 146

CONTRACT ADMINISTRATION

The engineer should ensure that the construction contract is administered in accordance with its terms. All too often, a contract is administered in a manner that is to some extent contrary to its terms. For example, payment dates may be overlooked; contract extras may proceed without written authorization required by the contract; parties may ignore notice requirements that were stipulated in the contract. The conduct of the parties during construction may bear little resemblance to the conduct that was contemplated when the contract was signed. Conceivably, the issue of equitable estoppel might arise; for example, if contract "extras" proceed without written authorizations, as contractually required, a party may be equitably estopped from denying that it had waived its contractual rights. It is better to avoid complicating the matter; this can best be achieved by closely administering the contract in accordance with its terms.

The engineer-administrator on a construction project should keep very detailed records. Thus the engineer will have evidence of the manner in which the contract was administered and the circumstances of the actual construction. The records should include a daily diary; carefully drafted minutes of meetings with contractors' representatives, owner, and engineer; and detailed notes of all developments, which might give rise to claims, during the construction. The engineer might be asked later to recollect events during construction; for example, in connection with subsequent delay and interference claims.

DRAWINGS AND SPECIFICATIONS

One of the most important services provided by the engineer on a construction project is the preparation of conceptual and detailed drawings and specifications to describe the work. The engineer's description of the work will form the basis of contractors' prices; and the engineer's client will rely upon the plans and specifications as well.

During construction, disputes often arise as a result of incomplete or ambiguous specifications. For example, where mechanical hardware — door locks, fasteners, and so on — is mentioned in the specifications; although quantity is specified, there may be no reference to a particular product or manufacturer. If the contractor installs hardware of poorer quality than that anticipated by the project owner, and if the construction contract has a stipulated-price or lump-sum basis, then a dispute may well arise. It is surprising how often an owner and a contractor will begin construction without detailed drawings and specifications; in an attempt to expedite construction, usually to avoid escalating costs or to take advantage of more favourable weather conditions. Such action might reflect the best intentions. But it can obviously lead to disputes about the true meaning and intent of the contract documents.

Detailed project drawings and specifications should include all required dimensions, units of measurement, quality of materials, and also product designations of parts and equipment. Product standards should include references to recognized testing authorities, such as the Canadian Standards Association ("CSA") or the American Society for Testing Materials ("ASTM"), wherever possible.

The importance of detailed specifications was illustrated in the 1979 decision in *Trident Construction Ltd.* v. *W.L. Wardrop and Assoc. et al.*[8] Justice Wilson, of the Manitoba Queen's Bench, displayed a sympathetic attitude to the contractor where the engineer had failed to specify a proper design. The Judge considered whether the likelihood of disaster might have been detected if the contractor had investigated or reworked the engineer's design:

> In the building trade in Manitoba not only does the owner rely upon a consulting engineer to prepare competent and reliable structural drawings for the erection of a building in accordance with those drawings, but also the contractors who were called upon to tender for those buildings on the basis of such drawings and specifications must also rely upon the accuracy and competency of these drawings. The normal practice is for the engineers to take months to research and prepare the plans and specifications involved in a project for which they are paid substantial sums and to give the contractors only a few weeks to prepare and to submit competitive bids for the construction of the building based on such plans and specifications. In this time frame it is impractical for any building contractor in Winnipeg to check out independently the structural design of buildings involved in the plans and specifications prior to bidding. During the few weeks open to him he is fully involved in doing take-offs and getting and collating prices. During that time he has to go

---

[8]   Supra, at page 38

through volumes of specifications and many sheets of plans to establish what materials, machinery and equipment he has to order, and to compute the labour involved in obtaining and procuring, fabricating, constructing and equipping such a plant. The task is made much more difficult by the series of late, extensive and complex addenda that are usually issued up to the very day on which tenders are called. The general contractor must also obtain bids from the various sub-contractors, and the sub-contractors must in turn obtain quotations from their various suppliers and sub-subcontractors, and the whole must be gathered together to make up the total bid of the general contractor within the bidding time allowed. There in fact is no time for anything else except the foregoing under the standard bidding procedures for a large project ... and the bidding represented no exception to the general rule.

THE TENDERING PROCESS

Where contractors submit competing bids, in the project tendering process, the acceptance of any one of these bids constitutes a contract — or the basis for the formation of a contract — between the successful contractor and the owner. The selection depends on the particular terms the owner has outlined in its tendering documentation. As the owner's agent, the engineer is usually involved in the preparation of tendering documentation.

The "Information to Tenderers" package should be tailored to the particular project. It will normally include a general description of the nature of the project; plans and specifications; information with respect to where more detailed data may be obtained (for example, a reference to soils reports); bid deposits or bid bonds that will be required at the time of submitting the tender; and the date, time, and place for submission of tenders. The package also normally includes the statement that the lowest or any tender will not necessarily be accepted by the owner.

The tendering issues identified in Chapter 16 should also be carefully considered in preparing the tender package and in administering the award of the contract. The Contract "A" complexities identified in Chapter 16 complicate the preparation of the tender package and "preventive" legal advice is typically advisable.

A tender is a contractor's offer to complete construction as described in the bid submission. Some tendering procedures are set up so that the owner's acceptance of the contractor's offer will constitute a contract between the owner and the contractor. It is important to include a form of the construction contract that the owner is prepared to sign as part of the documents delivered to the contractor prior to its bid submission. The engineer should advise the client of the need for preparing a contract form that is appropriately "tailored" to the project — a drafting process that is best left to a lawyer who can prepare the contract on the basis of the engineer's and the client's instructions.

Contractors who submit bids on construction projects and on equipment-supply contracts will often qualify the basis upon which their bids are submitted. They might, for example, take exception to terms and conditions included in the owner's form of contract. Such qualifications change the basis of the proposed contract. Until the owner is prepared to accept the contractor's revised terms and conditions, the formation of the contract will be delayed. Usually there is a straightforward business negotiation between the owner's and the contractor's representatives, the parties try to work out an agreeable basis for the contract so the work may proceed subject, of course, to avoiding breach of Contracts "A" as referenced in Chapter 16. The engineer might be engaged by the owner to assist in the negotiation process. The engineer should monitor closely the details of the negotiation and should document the final agreement between the parties. A well-drafted contract can then be finalized and signed. Where work proceeds before detailed terms and conditions are finalized, the potential for contract disputes increases. Whenever construction proceeds before contract documents have been finalized, the door is open to unnecessary headaches in contract administration.

The contract should be finalized and executed at the earliest possible opportunity. The contract provides the engineer-administrator with a defined basis upon which to proceed.

TYPES AND FORMS OF CONSTRUCTION CONTRACTS

Under common law, parties have the right to choose their contract terms and conditions; thus there is no prescribed format for construction contracts. There are, however, certain types of contractual arrangements and contract formats that are being used in the industry. Some of these formats have been reduced to "standard forms" that are widely accepted. However, the engineer should adopt a critical approach to the use of standard-form construction contracts (or the use of any standard-form contract). The parties should closely examine a proposed standard form to determine if it reasonably reflects the agreement they wish to express. Any modifications to the contract should be clearly and appropriately drafted (preferably by a lawyer) and incorporated into the wording of the contract prior to its execution.

Some examples of different types of contracts currently in use in the construction industry are listed below.

1. **Stipulated-Price or Lump-Sum Contract**   This type of contract gives the owner the benefit of knowing the total price that will have to be paid to the contractor for the completion of the construction (subject to additions or deductions to or from the work as the course of construction proceeds) in accordance with the terms of the contract documents. When the stipulated-price contract form is used, it is essential that detailed plans and specifications form the basis of the contractor's price. The engineer should carefully define and fully detail the work involved at the outset. If these details are provided after the price has been tendered, the contractor may claim that the work is beyond the scope of the contract as the contractor understood it at the time of bidding; the contractor may claim that additional compensation is warranted.

The stipulated-price contract may involve a considerable degree of risk for the contractor. For example, extra costs may be incurred if subsurface, dewatering, or weather problems arise. The contractor should set the initial price accordingly.

The stipulated-price contract does provide the owner with the advantage of a reasonable approximation of the total cost of the construction. But a word of warning: if the contractor has submitted an unrealistically low price for the work, the contractor may conceivably try to cut costs and look for opportunities to claim extra compensation. Insufficiently detailed specifications may well assist a contractor in such attempts.

2. **Unit-Price Contract**   This type of contract is often used for projects where it is difficult to predetermine quantities. For example, in an excavation project, the extent of varying subsurface conditions might not be determinable in advance. Bids are submitted on the basis of price per unit of item. In preparing specifications, the engineer should ensure that the quantity units on which prices will be based are clearly spelled out and should

also list alternatives. In excavation projects, for example, the engineer should request prices for removal of either rock or earth.

3. Cost-Plus Contracts

(i) **Cost Plus Percentage**   This type of contract provides compensation to the contractor for its costs incurred; as well, it provides a reasonable percentage to cover the contractor's overhead and profit. The cost-plus approach is often used on large-scale projects where there is not enough time to finalize detailed plans and specifications; to take the time may not be practicable in light of the urgency to proceed with construction. In large-scale projects, there is also the likelihood of changes in the work that the owner may wish to effect from time to time. Note that the cost-plus contract provides no incentive to the contractor to reduce costs: the contractor is being paid a percentage of the total construction costs. The engineer-administrator should closely monitor the progress of the work and the contractor's accounts in order to ensure that the work is performed at a reasonable cost. It is advisable for the owner to be expressly entitled to access all of the contractor's project records. This will facilitate the extensive, detailed administration that is usually necessary.

(ii) **Cost Plus Lump-Sum Fee**   This contract approach is much like the cost-plus-percentage arrangement. But the contractor does not receive a percentage of the project costs: instead, the contractor is paid a fixed amount. This may be to the owner's advantage. However, although the contractor receives no incentive to increase the total cost of the contract, there is also no incentive to reduce costs. As with the cost-plus-percentage arrangement, the engineer-administrator should closely monitor the work to ensure that the contractor is not being careless in coordinating and scrutinizing work that is performed by the various trades.

(iii) **Cost Plus Lump-Sum Fee plus Bonus**   In this type of arrangement, the contractor is provided with an incentive to reduce costs: for every dollar saved on an agreed estimated total cost, the contractor may receive an additional compensation in the form of an agreed-upon percentage of the saving. This approach makes good sense for the owner, as long as the agreed estimate is a reasonable one. Usually, the owner will expect the engineer to advise the owner on the reasonableness of the contractor's estimate. An inflated estimate would obviously increase the contractor's likelihood of benefiting from the bonus arrangement.

4. Guaranteed Maximum Price plus Bonus Like the cost-plus-bonus contract, the guaranteed-maximum-price contract incorporates an incentive to the contractor to effect savings on a cost-plus project. The contractor receives a fixed fee as well as an agreed-upon percentage of savings. The guaranteed-maximum-price contract offers the owner a further advantage over a basic cost-plus contract: the guaranteed-maximum-price feature. The amount of the guaranteed maximum price should be determined on the basis of detailed plans and specifications rather than on the basis of a reasonable estimate. Again, the owner usually expects the engineer to advise on the reasonableness of the contractor's maximum price.

5. Design-Build Contracts With this type of contract it is normally the contractor — rather than the owner — who arranges to obtain the necessary engineering design services, to finalize the design detail. The detailed design is often finalized as the construction work proceeds. Thus construction may proceed promptly; detailed plans and specifications for subsequent phases are finalized at a later date. The work often proceeds on a cost-plus-bonus basis; this appeals to owners who wish to avoid retaining professionals. In design-build contracts, the engineer is usually the agent of the contractor. But the contract normally entitles the owner to retain, at its option, its own engineering representative. The owner's engineer will double-check the sufficiency of the design services provided; the engineer will also be expected to ensure that construction proceeds in compliance with the agreed-upon plans and specifications.

PROJECT MANAGEMENT

The design-build contract is often used in connection with projects that are organized on a "project-management" basis. The owner usually enters into a contract with a project manager. The project manager acts as the owner's agent and acts on the owner's behalf to arrange for professional design services and to hire contractors to complete the construction. The project manager usually purports to have the experience and contacts necessary to facilitate and expedite the design, tendering, and construction stages. In return for acting on the owner's behalf, the project manager receives a fee; this fee is in addition to the professional design fee and the contracting cost, which the owner would incur in any event. The project manager simplifies the construction process for the owner; theoretically, the project manager's fee is the premium the owner is prepared to pay in order to receive the advantages of the project manager's experience, contacts, and administrative expertise. Hiring a project manager may, in fact, lead to overall cost savings on the project.

PRIME CONTRACT AND SUBCONTRACTS

Where the owner enters into a construction contract with a general contractor who in turn enters into subcontracts with various trade contractors, a contractual relationship, called "privity of contract," exists between the owner and the general contractor; privity of contract exists between the general contractor and each of its trade subcontractors. But no privity of contract exists between the owner and any of the subcontractors. The administration of the contract between the owner and the general contractor should be consistent with the administration of the contract between the general contractor and each of its subcontractors. To ensure consistency, the general terms and conditions of the head contract — between the owner and the general contractor — should be included in each of the subcontracts wherever applicable. If this consistency of terms and conditions is not achieved, there is no assurance that the general contractor will be capable of passing on to the subtrades the engineer-administrator's decisions. Inconsistency increases the likelihood of contract administration difficulties. For example, suppose mechanical extras are authorized and valued by the engineer pursuant to the terms of the head contract. And suppose the general contractor does not provide, in the mechanical subcontract, that extras are to be valued by the engineer. If the mechanical subcontractor disputes the value of the extras, unnecessary difficulties may well arise. The subcontractor may be entitled to claim an amount for the extras in excess of the engineer's valuation. The general contractor will then have a problem, which may conceivably adversely affect an otherwise smooth contract administration.

The engineer-administrator can promote smooth contract administration by ensuring that the general contractor is aware of the need for consistency in the general conditions of the head contract and the general conditions of each subcontract.

DELAY AND INTERFERENCE CLAIMS

Claims by contractors against owners for damages resulting from delays and interference caused by the owner's representative are not at all uncommon in the construction industry today. Claims on account of delay may result when enquiries from the contractor are not properly answered; the contractor may be required to delay construction until receiving a reply to his or her query. Claims of interference result when the works of subcontractors overlap or when a project is not well coordinated. The engineer-administrator should ensure that responses to contractors' enquiries are provided promptly and that the overall project is coordinated to minimize delay and interference claims.

COMPLIANCE WITH NOTICE PROVISIONS

The entitlement of contractors to time extensions or other relief for unforeseeable or excusable delays is usually conditioned upon the contractors giving written notice of delays within a contractually stipulated time.

As a general rule, the courts have not been very forgiving; they have strictly interpreted provisions of contracts requiring timely notice.

For example, in the case of *Corpex (1977) Inc.* v. *Canada*,[9] a decision of the Supreme Court of Canada, an excavation and construction contractor undertook, by contract with the Crown, to build a dam. After completing the project, the contractor brought an action claiming for the costs incurred due to mistaken information provided by the Crown's engineers regarding the nature of the soil, and for the costs resulting from delays and disruptions caused by the performance of work at the request of the owner's engineer in addition to that provided for in the plans and specifications, i.e., for "extras." The Supreme Court allowed the claim for the "extras," because there was no notice provision in the relevant contract clause. With regard to the costs due to the mistake regarding the nature of the soil, the contract allowed for reimbursement of additional expenses, conditioned on written notice of the claim being provided by the contractor within thirty days after encountering the soil conditions giving rise to the claim. Justice Beetz, writing for the Supreme Court of Canada, held that, since the contractor did not provide the requisite notice

[9]   [1982] 2 S.C.R. 643

but rather waited until completion of the contract, the contractor's claim could not succeed. Justice Beetz explained that the soil conditions reimbursement clauses provided benefit to both the contractor and the owner: the contractor was assured of recouping its additional costs if it complied with the notice provisions, and the owner could at that time have decided whether to continue with the project or to rescind the contract. However, if the contractor waited until the work was complete before claiming for the reimbursement, the contractor then deprived the owner of the contract rescission benefits that the contract expressly provided for the owner.

The lower courts have followed the lead of the Supreme Court of Canada in strictly construing the notice provisions of contracts.

In the case of *Doyle Construction Co.* v. *Carling O'Keefe Breweries of Canada Ltd.*,[10] the British Columbia Court of Appeal considered a contractor's claim for "impact costs" incurred by the contractor as a result of the owner's delays, interference, and changes of the work sequence. The contractor contended that it could not ascertain the cumulative impact cost until the completion of the construction, when the claim was finally made.

Along with other reasons for dismissing the contractor's appeal, the appellate court found that the trial judge had correctly rejected the contractor's claim because the contractor had failed to comply with the terms of the contract requiring the claim for damages to be made in writing within a reasonable time after the first observance of such damages.

In the case of *Acme Masonry Ltd.* v. *Byrd Construction Ltd.*,[11] a subcontractor entered into a subcontract with the defendant general contractor who was to construct a court house. As a result of labour unrest and the unavailability of certain materials, the performance of the subcontract was delayed for several months beyond the original schedule. The subcontractor accepted the revised schedule but gave notice in writing that it intended to make a claim beyond the amount of the subcontract as a result of the delays. By the terms of the subcontract, the subcontractor was required to set out its claim fully in writing within a reasonable time after the first observance of the damages claim, and not later than the time of issuance of the final certificate.

The British Columbia Court of Appeal held that, notwithstanding that the subcontractor had put the contractor on notice that it

---

[10] (1988), 23 B.C.L.R. (2d) 89 (C.A.)
[11] (1986), 20 C.L.R. 228 (B.C.C.A.)

would be claiming for the additional costs, the subcontractor's claim was "statute barred" as it had failed to submit the detailed claim before the date of the final certificate. "Notice of the notice" was not sufficient; the notice had to be provided on time.

However, notwithstanding these cases of strict judicial construction of notice provisions, a 1987 decision of the British Columbia Supreme Court appears to have broadened the court's interpretation of what constitutes "notice." In the case of *W.A. Stevenson Construction (Western) Ltd.* v. *Metro Canada Ltd.*,[12] the plaintiff was the general contractor for the construction of a portion of the Vancouver Light Rapid Transit System. The construction contract with the defendant developer provided for six milestone completion dates within the overall contract time and an agreed completion date for the whole contract. The contractor brought an action against the developer claiming $4.6 million in damages for breach of contract, on the basis that it had been delayed by not being given full access to the construction site and then forced to accelerate construction to maintain the tight schedule.

The court found in favour of the contractor. The court found as a question of fact that the developer had failed to provide the contractor with the unobstructed workspace called for by the contract; failed to provide time extensions to which the contractor was entitled due to the obstructed workspace, inclement weather, and labour problems; and failed to provide timely and accurate information concerning factors adversely affecting the contractor's planning. The effect of these failures was an interference with the contractor's efficiency, disruption of the coordination of its construction operations, and a loss of its productivity, all resulting in a loss of the opportunity of making the original profit.

In a departure from a strict construction of notice provisions, the court held that the contract terms requiring the contractor's timely written notice of delays and extra claims, or the developer's written waiver of its rights to require such formal notice, did not bar the contractor's recovery, despite the lack of any formal notice or written waiver. The court determined that "constructive notice" of the contractor's problems was provided to the developer by the contractor in the form of miscellaneous written communications and detailed minutes of meetings. The court observed:

[12] (1987), 27 C.L.R. 113

A reading of the minutes is very revealing: they were obviously regarded by everyone as a method of formally communicating their concerns to the other party.

Thus, the intention of formal communication between the parties was sufficient to constitute notice, notwithstanding non-compliance with the formal procedure of written notice. The *Stevenson* decision may open the door for other varieties of "constructive notice" to assist contractors in advancing the otherwise often difficult argument that, in the circumstances of a particular project, compliance with strict notice requirements should not be a prerequisite to recovery.

DEPARTURES FROM TRADITIONAL CONTRACTING APPROACHES

A further caution about the use of "standard-form contracts" arises because of the approach that may be adopted, particularly on larger projects, to depart from the "traditional" contract arrangements. For example, the popular "fast-track" or phased-construction approach, where construction proceeds on the designed portions of a project while other portions are still being designed, often involves the services of a construction manager who coordinates the project and also might do some construction work. In such a situation, it isn't possible to achieve a traditionally structured single-stipulated-price contract. Such special projects require specially drafted contracts.

The use of multiple-prime contractors ("multi-primes") on projects is another structuring approach. Again, the traditional standard-form contracts are not appropriate. Multi-primes may, for example, be used by developers who see themselves as general contractors and who parcel out the work of the various trades on individual prime-contract arrangements. As construction managers or project managers, engineers may be required to coordinate and schedule multi-primes. The project contracts must be appropriately negotiated and should realistically describe the responsibilities and protect the interests of the various participants in the construction project.

As professionals in the construction industry find new approaches such as BOT projects and public/private partnerships to respond to project requirements, the need for "hybrid" and a variety of special contracts follows. As more fully explained in

Chapter 25 "BOT" is an acronym for "build, operate, transfer," and it is commonly used on projects where the contractor not only designs and constructs the project but is also responsible for its financing, in whole or in part. The contractor operates the project over a fixed period and receives the benefit of tolls or other revenues of operation, and is typically required to transfer the project to the government or private "owner" at the end of the operating term. Engineers play a key role on projects implementing these new approaches and should seek the advice of an experienced lawyer before finalizing such contracts. Contracts that are prepared by contract administrators and that do not receive appropriate attention from legal counsel may well create unnecessary complications when difficulties and claims later arise.

THE SHIFT TO INFRASTRUCTURE

The 1980s and 1990s have been periods of substantial change within the construction industry in Canada. With the arrival of the recession in the early 1990s, the booming development marketplace of the late 1980s saw an abrupt reversal. Those who were able to weather the recessionary storm witnessed a major change in the industry. In general terms, the opportunity focus within Canada shifted dramatically away from the traditional high-rise and similar commercial development to infrastructure projects — a shift that has obviously been experienced elsewhere in North America and in many countries around the world. Governments at all levels within Canada are supporting these infrastructure initiatives. Current primary areas of infrastructure development and planning in Canada include transportation projects, such as toll highways, subway extensions, airports, and water and sewage facilities. These Canadian infrastructure opportunities are attracting a great deal of interest from U.S. and other international contractors.

As further explained in this book, the North American Free Trade Agreement ("NAFTA") substantially increases the ability of North American contractors to provide construction services in Canada, the United States, and Mexico, without discriminatory restrictions, particularly in the area of government procurement contracts for construction services.

Indicative of government support for infrastructure improvements are the federal government's Canada Infrastructure Works

Program and the Community Economic Development Act of the Province of Ontario.

Canada Infrastructure Works

Responding to the importance of the infrastructure focus, Canada's federal government initiated a special $6 billion national program, known as Canada Infrastructure Works, designed to renew and upgrade the quality of Canada's physical infrastructure. Its costs are to be shared by the federal, provincial, and municipal governments.

As already noted, the infrastructure projects range from traditional facilities such as roads, highways, bridges, water systems and sewers to municipal and educational buildings, waste management, information technology, and recreation and tourism facilities.

In addition to infrastructure renewal, the program placed emphasis on the provision of timely and effective employment creation and skills development; the improvement of national, provincial, and local economic competitiveness; and the promotion of improved environmental quality.

The Community Economic Development Act

Ontario's Community Economic Development Act (the "Act") is illustrative of a statute that allows municipalities to participate in public/private infrastructure projects, a very popular focus in Canada.

Among other provisions, the Act provides for the establishment of community investment share corporations and community loan fund corporations and for the payment of grants, as well as for limited provincial guarantees for these corporations.

One of the most interesting provisions of the Act permits a council of a municipality to enter into agreements for the provision of "municipal capital facilities" by any person. Such an agreement allows for the lease, operation, or maintenance of those facilities by a private sector operator, thereby providing an avenue receptive to "privatization."

In addition, the council of a municipality may provide financial or other assistance at less than fair market value or at no cost to the person who has entered into an agreement for the provision of municipal capital facilities. However, assistance may be only in respect to the provision, lease, operation, or maintenance of the facilities that are the subject of the agreement.

The definition of what constitutes a "municipal capital facility" is critical and it can include the following:

- municipal roads, highways, and bridges
- municipal local improvements and public utilities, except facilities for the generation of electricity
- municipal facilities related to the provision of telecommunications, transit, and transportation systems
- municipal facilities for water, sewers, sewage, drainage, and flood control
- municipal facilities for the collection and management of waste and garbage
- municipal facilities related to the provision of social and health services, including homes under the Homes for the Aged and Rest Homes Act
- municipal facilities for public libraries, and
- municipal facilities used for cultural, recreational, or tourist purposes

A municipality may also exempt from taxation, for municipal and school tax purposes, land on which municipal capital facilities are, or will be, located. A similar full or partial exemption may be granted for development charges.

As a result of the revisions to the Municipal Act, municipalities are now allowed to enter into arrangements with the private sector to provide for the design, construction, financing, operation, and maintenance of municipal capital facilities.

Public/private infrastructure partnerships are likely to continue playing a major role in Canadian infrastructure projects on a scale that should provide substantial opportunities for the engineering community. Toll highways and water and sewage projects are likely examples.

SIMILARITIES BETWEEN CANADIAN AND AMERICAN CONSTRUCTION INDUSTRIES

The Canadian construction industry has many similarities to the American construction industry. This is not surprising, given the close ties between the countries and the long history of participation by American contractors in Canada. Approaches to contract documentation, for example, often reflect substantial similarities as do bonding and insurance arrangements and claims experience on construction projects. ADR, BOT, and public/private partnerships are "leading edge" developments in both countries.

STUDY QUESTIONS

The following hypothetical cases and commentaries are included for illustrative study purposes.

1. In some construction contracts, an engineer is authorized to be the sole judge of the performance of work by the contractor. Where such a provision is stated, is it possible that the provision will not be enforceable on account of the manner in which the engineer performs his or her duties? Explain.

Commentary: 1. The key here is whether or not the engineer is biased in making decisions. If he or she can be shown to be biased (and this will not be easily proven, typically) then the provision that the engineer's decision is binding will not be enforced.

2. Usually, an engineer on a construction project is authorized to inspect the construction in order to ensure that the work proceeds in accordance with plans and specifications. Comment on the desirability of an engineer actually instructing the contractor on work methods and construction procedures that the contractor should employ in carrying out the work.

Commentary: 2. The engineer must be careful not to interfere with the contractor's contractual rights to choose work methods and procedures. If the engineer does so, the contractor may claim that it was delayed by the engineer. If the engineer interferes negligently, perhaps by giving negligent advice, a tort claim may arise against the engineer.

3. Briefly describe how a bonus feature may be included in a guaranteed maximum price construction contract to the advantage of both the owner and the contractor.

Commentary: 3. Refer to chapter materials relating to sharing of savings on guaranteed maximum price (GMP) contracts providing an incentive to the contractor by way of sharing with the owner in any amount by which the actual cost of construction is less than the GMP.

CHAPTER TWENTY-FIVE

RISKS IN CONSTRUCTION

Construction has long been regarded as a high risk industry. That being the case, it is advisable for each participant involved on a construction project to carefully address the risks that may arise in the circumstances of each particular project with a view to minimizing their impact. An enlightened approach to dealing with risks at the outset by project participants can lead to lower project costs and lessen the likelihood of claims and disputes. But to achieve such an objective requires a careful and considered approach to risk identification, risk assessment, risk allocation, and contract administration. To maximize the benefits of the process to the advantage of various contracting parties calls for cooperative attitudes and the endorsement of the principle of reasonable and equitable risk allocation. Owners who insist on draconian contract provisions such as "no liability for delay" clauses and the imposition of harsh subsurface provisions, for example, may well wind up with much higher prices than would otherwise have been bid. Contractors who pursue additional compensation through aggressively pursuing extras and claims (often in circumstances where market conditions may have forced them to accept unreasonably allocated risks) similarly complicate the project outcome. The result of aggressive and adversarial contracting attitudes has too often been protracted, expensive, and disruptive litigation or arbitration.

This chapter addresses various aspects of risk assessment and allocation on construction projects, particularly with respect to project planning and structuring considerations as well as the contractual allocation of risks. The importance of a contract administration process that ensures compliance with the contract documents and hence the implementation of the project on a basis consistent with the agreed allocation of risks is also emphasized.

Some risks in construction are relatively straightforward and readily assumed by project participants. Others, such as "no damages for delay" provisions and the harsh subsurface conditions provisions already mentioned, may well lead to controversy and set the stage for adversarial and acrimonious project relationships. Unfortunately, owners and contractors alike have pursued aggressive positions to preclude problems encountered by themselves or others in the past. Fortunately, more and more emphasis is being placed on the benefits of "partnering" on projects in order to develop more cooperative team approaches to construction. Increased cooperation should assist in achieving as smooth a contract administration process as possible, all in the interests of minimizing or avoiding the undesirable cost and disruptive impacts of project disputes.

The risk assessment and allocation process needs to be carried out on a project by project basis. Some specific project risks will be identified in this chapter and many of them will be seen to be typical traditional risks in construction. However, the nature of risks can vary significantly from project to project. International projects, for example, often involve additional risks relating to currency exchange, convertibility of foreign currencies, and inflation risks. These factors may have a very significant bearing on, for example, a build-operate-transfer ("BOT") project in a developing country financed with Canadian or U.S. dollars, where the stream of future revenue earnings to the contractor is based on the local currency. If the local currency is unstable and future operating periods generate inflated and devalued currency returns, the actual project revenues may fall far short of the projected returns pursuant to the economic feasibility analysis for the project.

International projects typically often also add political risk considerations. Political risk insurance should be considered. It may or may not be available or affordable, but it is advisable to consider political risk very carefully, particularly in less stable countries.

In addition, the popular trend to BOT and public/private partnership projects exposes the traditional contractor to many new "operational" risks; for example, in the operation of a toll highway or in the operation of a co-generation power plant. BOT projects require participants to put in place operating contract documents that will address events and circumstances that may not arise for many years into the future. In such circumstances, careful contract wording is very important in anticipating how to deal with future events, particularly given the changes in project personnel that will undoubtedly occur. The risk of contract interpretation difficulties is one that always warrants a careful and considered approach and the early involvement of legal counsel to draft carefully coordinated agreements on these complex projects.

PROJECT STRUCTURING

As experienced participants are aware, it is an important aspect of the risk assessment process on each project to analyze the nature of the contractual relationships between various project participants. For example, depending upon the particular circumstances of each project, it may be considered advisable from the contractor's perspective to provide services as a construction manager rather than as a general contractor. Doing so typically enables the construction manager to avoid contractual responsibility for the performance of trades that would otherwise be performing as subcontractors on the project. Another example of a departure from a traditional approach is where the project architect does not assume contractual responsibility for the normal structural, mechanical, and electrical engineering services.

A comparison of the "traditional" project structure with a construction management "multi-prime" structure is shown as follows:

TRADITIONAL PROJECT STRUCTURE

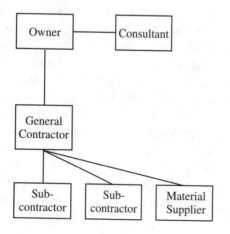

HYBRID STRUCTURE
CONSTRUCTION MANAGEMENT
(Multi-Primes)

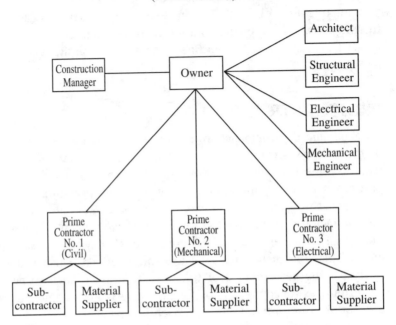

Of course, whenever the burden of risk shifts in a construction project as a result of contractual relationships, it is in the interests of each contracting party to make sure that it understands the potential extent to which particular contractual relationships on a project may result in limiting liability. For example, referring again to the examples already noted, it may be that the project owner prefers to contract with a single general contractor on a project rather than to enter into a construction management arrangement that may result in the owner entering into a multiplicity of contracts with various trades (the "multi-prime structure"). Similarly, the project owner may also prefer to contractually obligate its architect for engineering services, rather than to enter into agreements with a variety of consultants on a project. The advisability of a particular project structure should be considered carefully in the circumstances of each project and it is extremely important that it be carefully considered at as early a stage as possible. Prior to assembly of instructions to bidders, owners should carefully assess any recommended project structure in order to be satisfied that the resultant contractual responsibilities will be consistent with the owner's expectations as far as risk assumption is concerned. The owner's engineer should provide guidance in that assessment process, particularly where the engineer's client may be relatively inexperienced in project matters.

Similarly, other hybrid structures are generated by special project circumstances such as design-build projects and BOT and public/private infrastructure projects.

As emphasized in this chapter, risks flowing from contractual relationships need to be carefully assessed in the circumstances of each project. In a BOT project, for example, the additional scope of the contractor's typical responsibilities for financing, designing, building, operating, and ultimately transferring the project usually generate a significant number of project contracts well beyond the scope of even an experienced engineer's expertise.

CONTRACT FORMS

A major factor that may bear upon the decision of an owner and its advisors in determining project structure will be the contract forms that may be preferred for use. In the Canadian construction industry the standard form contract documents published by the Canadian Construction Documents Committee ("CCDC") are widely used. As described in Chapter 24, these CCDC forms have evolved over the years and are the product of a committee composed of a variety of construction industry participants, namely representatives appointed by each of The Association of Consulting Engineers of Canada, The Canadian Construction Association, The Canadian Council of Professional Engineers, Construction Specifications Canada, and The Royal Architectural Institute of Canada. As such, the CCDC documents have the benefit of perspectives of contractors and of a broad range of consultants familiar with Canadian construction projects. In addition, such consultants typically act for owners on construction projects. However, strictly speaking, owners are not represented on the Canadian Construction Documents Committee.

The CCDC standard form construction contracts referred to are between owner and contractor. Although not a party to these contracts, the engineer as the consultant plays a very significant role as contract administrator with extensive duties and responsibilities as set out in the contract documents. Such contract administration responsibilities, together with the consultants' design responsibilities, give rise to the consultants' personal interests also being affected by the CCDC contract documents or, for that matter, by any construction contract setting out functions to be performed by a third party consultant. Accordingly, although not a party to such a construction contract, the engineer as consultant has a personal interest in such a document insofar as the consultant's duties are reflected and described in the construction contract (and typically and advisedly cross-referenced in the client-consultant agreement).

Also as previously referenced, the CCDC standard contracts include provisions to address the interests of both the contractor and the owner and include a comprehensive range of general conditions typical to construction contracts. However, particular project circumstances may well necessitate "tailoring" of the standard form contract, often as a result of preferences of experienced owners and developers who may prefer a particular or a more detailed approach to certain aspects of the CCDC standard form general conditions. This approach may be more aggressively in the owner's interests than that reflected in the CCDC forms. Similarly, contractors may choose to pursue more advantageous provisions in subcontracts than provided in CCDC standard forms. It is very important for all parties to review contract forms carefully. Understanding the potential significance of proposed contract terms is a vital part of the project risk assessment process. It may be that some risks can be appropriately shifted and covered, from a business perspective, by contract pricing adjustments.

It is also important to bear in mind that the CCDC standard form contracts are based on the traditional project structure where the owner enters into a contract with a consultant and the owner also enters into a contract with a general contractor. Accordingly, tailoring will also be necessary where the project structure is not reflective of this "traditional approach." For example, the CCDC standard form contracts would require modification in circumstances where the owner enters into a multi-prime contractual arrangement involving a construction manager. In addition, there are special pricing arrangements, such as cost-plus-percentage or fixed fee contracts that include a guaranteed maximum price provision (perhaps also including a sharing of savings formula to the extent that actual costs of construction are less than the guaranteed maximum price). These arrangements require modifications to the pricing provisions if the CCDC standard form contract is used as the basis for the contract, as it often is.

Not surprisingly, construction contracts contain a variety of provisions limiting, to some extent, potential liabilities of the contracting parties, either with respect to limitation periods (i.e., time periods during which lawsuits may be commenced) or the monetary limits on claims that may be made. For example, owners should note that GC 12.1.2 of the 1994 CCDC 2 standard contract forms provides:

> 12.1.2   The obligation of the *Contractor* to indemnify hereunder shall be limited to two million dollars per occurrence from the commencement of the *Work* until *Substantial Performance* of the *Work* and thereafter to an aggregate limit of two million dollars.

GC 12.1.2 is obviously a clause designed for contractors and is typically deleted or modified (by raising the amount of coverage) by owners who are aware of it. Its inclusion in the CCDC standard form, however, causes many owners to regard the use of the CCDC forms with scepticism.

CONTRACTUAL ALLOCATION OF RISK

In the pursuit of fair and even-handed contractual allocation of risks, a helpful principle to bear in mind is the reasonableness of the risk in question being assumed by the contracting party that is in the best position to assess and manage that risk. Again, this is a process that should be undertaken in the circumstances of each project. With that concept in mind, this chapter considers the basis upon which some specific risks are allocated between the "traditional" owner and contractor, assuming, for reference and example purposes, the use of the 1994 CCDC 2 stipulated price contract form.

PROJECT FINANCING RISK

GC 5.1 of the CCDC form is intended to provide information and comfort to the contractor with respect to the owner's ability to fund the project, and thereby reduce the contractor's risk in that regard, by providing:

GC 5.1 FINANCING INFORMATION REQUIRED
OF THE OWNER

5.1.1   The *Owner* shall, at the request of the *Contractor*, prior to execution of the Agreement, and/or promptly from time to time thereafter, furnish to the *Contractor* reasonable evidence that financial arrangements have been made to fulfil the *Owner's* obligations under the *Contract*.

5.1.2   The *Owner* shall notify the *Contractor* in writing of any material change in the *Owner's* financial arrangements during the performance of the *Contract*.

RISK OF CONCEALED OR UNKNOWN CONDITIONS

GC 6.4 provides a basis upon which a change order can be issued in the event of the discovery of concealed or unknown conditions, including subsurface or otherwise concealed physical conditions. GC 6.4 provides:

GC 6.4 CONCEALED OR UNKNOWN CONDITIONS

6.4.1   If the *Owner* or the *Contractor* discover conditions of the *Place of the Work* which are:

.1 subsurface or otherwise concealed physical conditions which existed before the commencement of the *Work* which differ materially from those indicated in the *Contract Documents* or

.2 physical conditions of a nature which differ materially from those ordinarily found to exist and generally recognized as inherent in construction activities of the character provided for in the *Contract Documents*

then the observing party shall notify the other party in writing before conditions are disturbed and in no event later than 5 *Working Days* after first observance of the conditions.

6.4.2   The *Consultant* will promptly investigate such conditions and make a finding. If the finding is that the conditions differ materially and this would cause an increase or decrease in the *Contractor's* cost or time to perform the

Work, the *Consultant*, with the *Owner's* approval, shall issue appropriate instructions for a change in the *Work* as provided in GC 6.2 - CHANGE ORDER or GC 6.3 - CHANGE DIRECTIVE.

6.4.3   If the *Consultant* finds that the conditions at the *Place of the Work* are not materially different or that no change in the *Contract Price* or the *Contract Time* is justified, the *Consultant* shall report the reasons for this finding to the *Owner* and the *Contractor* in writing.

RISK OF DELAYS

The delay provisions of the CCDC standard form contemplate that the contractor will be entitled to both reimbursement and extension of time in the event of a delay caused by an act or omission of the owner or the consultant under the contract, and also in the event of delays caused by a stop work order issued by a court or other public authority, provided that the order was not issued as a result of the fault of the contractor.

However, under the *force majeure* provisions that are typically inclusive of any cause beyond the contractor's control (for example, labour disputes, strikes, lock outs, fire, unusual delay by common carriers, or unavoidable casualties), the contract time is to be appropriately extended but the contractor is not entitled to payment for costs incurred unless such delays result from actions by the owner. Precluding the recovery of additional payment to the contractor in such circumstances is consistent with the concept that neither party is in a position to control these risks.

THE RISK OF TOXIC AND HAZARDOUS SUBSTANCES AND MATERIALS

The CCDC standard form provides for the owner to assume the risk arising out of toxic or hazardous substances or materials on the construction site prior to the contractor commencing the work. This provision also provides for a procedure should toxic or hazardous substances be encountered. GC 9.3 provides, in part:

GC 9.3 TOXIC AND HAZARDOUS SUBSTANCES AND MATERIALS

9.3.1   For the purposes of applicable environmental legislation, the *Owner* shall be deemed to have control and management of the *Place of the Work* with respect to existing conditions.

9.3.2   Prior to the *Contractor* commencing the *Work*, the *Owner* shall

.1   take all reasonable steps to determine whether any toxic or hazardous substances or materials are present at the *Place of the Work*, and

.2   provide the *Consultant* and the *Contractor* with a written list of any such substances and materials.

9.3.3   The *Owner* shall take all reasonable steps to ensure that no person suffers injury, sickness, or death and that no property is injured or destroyed as a result of exposure to, or the presence of, toxic or hazardous substances or materials which were at the *Place of the Work* prior to the *Contractor* commencing the *Work*.

9.3.4   Unless the *Contract* expressly provides otherwise, the *Owner* shall be responsible for taking all necessary steps, in accordance with legal requirements, to dispose of, store or otherwise render harmless toxic or hazardous substances or materials which were present at the *Place of the Work* prior to the *Contractor* commencing the *Work*.

9.3.5   If the *Contractor*

.1   encounters toxic or hazardous substances or materials at the *Place of the Work*, or

.2   has reasonable grounds to believe that toxic or hazardous substances or materials are present at the *Place of the Work*,

which were not disclosed by the *Owner*, as required in paragraph 9.3.2, or which were disclosed but have not been dealt with as required under paragraph 9.3.4, the *Contractor* shall

.3   take all reasonable steps, including stopping the *Work*, to ensure that no person suffers injury, sickness, or death and that no property is injured or destroyed as a result of exposure to or the presence of the substances or materials, and

.4   immediately report the circumstances to the *Consultant* and the *Owner* in writing.

9.3.6   If the *Contractor* is delayed in performing the *Work* or incurs additional costs as a result of taking steps required under paragraph 9.3.5.3, the *Contract Time* shall be extended for such reasonable time as the *Consultant* may recommend in consultation with the *Contractor* and the *Contractor* shall be reimbursed for reasonable costs incurred as a result of the delay and as a result of taking those steps.

. . .

9.3.8   The *Owner* shall indemnify and hold harmless the *Contractor*, the *Consultant,* their agents and employees from and against claims, demands, losses, costs, damages, actions, suits, proceedings arising out of or resulting from exposure to, or the presence of, toxic or hazardous substances or materials which were at the *Place of the Work* prior to the *Contractor* commencing the *Work*. . . .

THE RISK OF CHANGES IN GOVERNING REGULATIONS

The CCDC form provides that any increase or decrease in cost to the contractor due to changes in taxes and duties after the time of bid closing shall increase or decrease the contractor's price accordingly. Similarly, the CCDC form contemplates the issuance of change orders where laws, regulations, or codes change subsequent to the date of bid closing, again consistent with the concept that the contractor should not be responsible to bear the risk of increased costs resulting from subsequent changes in the law.

CONSTRUCTION SAFETY RISKS

The CCDC form provides that the contractor shall be solely responsible for construction safety at the place of the work; for compliance with the rules, regulations, and practices required by the applicable construction health and safety legislation; and for initiating, maintaining, and supervising all safety precautions and programs in connection with the performance of the work. However, the contractor's obligations in that regard are made subject to responsibilities of the owner with respect to any separate contracts awarded for other parts of the project or when work is performed by the owner's own forces.

DISPUTE RESOLUTION RISKS

The CCDC form attempts to minimize or avoid the risk of prolonged construction litigation or arbitration by including a mandatory sequential approach to negotiation, mediation, and arbitration of disputes. The purpose of the mandatory negotiation and mediation provisions is to bring disputing parties to the settlement table to determine if their differences can be resolved without proceeding to arbitration or litigation, sooner than might otherwise be the case. If still unresolved, arbitration is mandatory if either party elects to proceed to arbitration, otherwise the parties may refer an unresolved dispute to the courts.

RISK OF PROCEEDING WITH CHANGES WITHOUT FINAL AGREEMENT ON PRICE AND TIME ADJUSTMENTS

The CCDC form protects the contractor from the risk of proceeding without a fully signed change order evidencing agreement on the value of a change or the valuation method. A separate administrative document in the form of a change directive will suffice where the owner requires the contractor to proceed prior to agreement upon the adjustment in contract price and contract time being resolved.

RISK SUMMARY

The above discussion of the CCDC approach with respect to these various risks is intended to be illustrative of the risk allocation process, premised on the principle that the party that is in the best position to assess and manage any particular risk should assume that risk. The reasonableness of that presumption with respect to any particular risk, however, should be tested in the circumstances of the particular project.

Perhaps one of the best examples of a risk assumed by the contractor that is likely to result in a higher contract price to the owner is the risk of concealed or unknown conditions. A contractor faced with that risk should include a contingency or "risk premium" in its bid price. Quantifying that contingency is itself an obviously difficult process. That difficulty may well lead to a very high contingency subject to the reality of marketplace restrictions on the contractor.

Without attempting to provide an exhaustive or definitive list of risks, some examples of other construction risks include:

- the availability of sufficient labour forces in the area of the project to carry out the construction;
- the responsibility to obtain various building permits and approvals of regulatory authorities;
- the risk of access to the site being suitable for the project;
- the sufficiency of the drawings and specifications;
- the likelihood of adverse weather conditions;
- the risk of cost escalations during the course of the project;
- the likelihood of delays in obtaining approvals from the owner, should approvals be required during the course of the work;
- risks relating to owner supplied materials and equipment;
- risks relating to the impact of environmental assessments; and
- risks relating to archaeological discoveries during the performance of the work.

One of the most important determinations in the project process is the selection of the other contracting party. Awarding the contract to the contractor with the lowest price may not necessarily be advisable in all circumstances. Minimizing risk in this regard through prequalification procedures is strongly recommended. Similarly, from the contractor's perspective, the extent to which the owner and its personnel may be inexperienced in the construction process should be taken into account. If the owner is not familiar with construction (and the decision-making process involved) or is likely to slow the process down as a result of taking too bureaucratic an approach, the contractor should factor those risks into its project plans. It may be advisable to provide for an appropriate contingency to respond to those concerns.

SECURITY FOR RISKS

Given the impossibility of controlling all aspects of risks, it is advisable for project participants to also look to appropriate security instruments to provide protection.

Performance bonds are available to provide a contractual basis upon which the owner can look to a surety company in the event that its contractor fails to perform or is adjudged bankrupt or a receiver is appointed on account of its insolvency. Labour and material payment bonds set out the contractual basis upon which a surety company is obligated to provide funds if the general contractor fails to pay its subcontractors or material suppliers.

In order to protect itself against lien claims from unpaid project participants it is important that the owner complies with the holdback and other requirements of applicable provincial construction lien legislation. Special bonding coverage, in the form of a holdback release bond, can be used as security in circumstances where an owner is prepared to release holdback funds prior to the expiry of lien rights. Prior to doing so, it is important that the nature and sufficiency of the holdback release bond is carefully considered and legal advice appropriately taken.

Liability and property insurance protection, carefully tailored to respond to the circumstances of each particular project, should be obtained on each project. Where design risks are involved, as in design-build or client-consultant agreements, professional liability insurance coverage should be carefully considered, particularly in terms of amount and term of coverage.

Other security approaches taken to protect against performance risks include unconditional or conditional letters of credit under which payments can be demanded in the event of contract defaults as specified. On international contracts, these instruments are often referred to as performance guarantees or, in fact, as performance bonds.

Another approach commonly taken to secure contract performance, and thereby minimize risks associated with performance, is through the use of liquidated damage provisions requiring the contractor to make payment in prescribed amounts in the event of nonperformance under the contract. When drafting the liquidated damages provision it is important to bear in mind that Canadian common-law courts will not enforce penalty provisions as a matter of policy. Accordingly, the use of the term "penalty clause" should be avoided in contract drafting and, to be enforceable, the amount of the liquidated damages should represent a genuine pre-estimate of the damages resulting from such nonperformance. Accordingly, particular care should be taken in calculating the basis for the amount of such liquidated damages and in wording the clause.

THE RISK OF UNENFORCEABILITY OF CLAUSES LIMITING LIABILITY

Contract provisions limiting liability must also be carefully drafted to clearly express the intentions of the parties, because a court, when interpreting contracts, will look to the specific wording in the process of assessing the intention of the parties.

Not surprisingly, clauses limiting liability are strictly construed by our courts. An ambiguous clause capable of more than one reasonable interpretation can be attacked on the basis of the rule of *contra proferentum*. That is, the clause will be interpreted against the party that drafted the ambiguous limiting clause. Accordingly, provisions limiting liability should be carefully and clearly expressed in order that the party relying on the clause can demonstrate that the loss or damage in question falls within the scope of the provision.

In order to appreciate the approach now taken by our courts to the enforceability of clauses limiting liability, it is also important to be aware of case law since 1970 dealing with the "doctrine of fundamental breach," as described in Chapter 20.

RISKS RELATING TO LIMITATION PERIODS

The issue of limiting liability has long been a very major one for contractors and other participants in the construction industry. The enforceability of exemption clauses limiting liability has been reviewed in Chapter 20, particularly the *Hunter Engineering* decision. Whenever the opportunity arises to limit your liability, it is advisable to do so, including also the time period during which claims may be made. The limitation period issues are discussed in Chapter 5.

From a construction contract perspective, there are essentially two general conditions in the CCDC standard forms that relate to the length of time during which the contractor may be liable in the common-law provinces. They relate to warranty (GC 12.3) and to a longer period of six years with respect to "substantial defects or deficiencies" (GC 12.2).

GC 12.3 of the CCDC form deals with warranty and provides:

GC 12.3 WARRANTY

12.3.1   The warranty period with regard to the *Contract* is one year from the date of *Substantial Performance of the Work* or those periods specified in the *Contract Documents* for certain portions of the *Work* or *Products*.

12.3.2   The *Contractor* shall be responsible for the proper performance of the *Work* to the extent that the design and *Contract Documents* permit such performance.

12.3.3   Except for the provisions of paragraph 12.3.6 and subject to paragraph 12.3.2, the *Contractor* shall correct promptly, at the *Contractor's* expense, defects or deficiencies in the *Work* which appear prior to and during the warranty periods specified in the *Contract Documents*.

12.3.4   The *Owner*, through the *Consultant*, shall promptly give the *Contractor* notice in writing of observed defects and deficiencies that occur during the warranty period.

12.3.5   The *Contractor* shall correct or pay for damage resulting from corrections made under the requirements of paragraph 12.3.3.

12.3.6   The *Contractor* shall be responsible for obtaining *Product* warranties in excess of one year on behalf of the *Owner* from the manufacturer. These *Product* warranties shall be issued by the manufacturer to the benefit of the *Owner*.

GC 12.2.1 of the CCDC forms also includes "substantial defects or deficiencies" beyond warranty items and provides, in part:

12.2.1   Waiver of Claims by *Owner*

As of the final certificate for payment, the *Owner* expressly waives and releases the *Contractor* from all claims against the *Contractor* including without limitation those that might arise from the negligence or breach of contract by the *Contractor* except one or more of the following:

.1   those made in writing prior to the date of the final certificate for payment and still unsettled;

.2 those arising from the provisions of GC 12.1 - INDEMNIFICATION or GC 12.3 - WARRANTY;

.3 those arising from the provisions of paragraph 9.3.5 of GC 9.3 - TOXIC AND HAZARDOUS SUBSTANCES AND MATERIALS and those arising from the *Contractor* bringing or introducing any toxic or hazardous substances and materials to the *Place of the Work* after the *Contractor* commences the *Work*.

. . .

.4 those made in writing within a period of 6 years from the date of *Substantial Performance of the Work*, as set out in the certificate of *Substantial Performance of the Work*, or within such shorter period as may be prescribed by any limitation statute of the province or territory of the *Place of the Work* and those arising from any liability of the *Contractor* for damages resulting from the *Contractor's* performance of the *Contract* with respect to substantial defects or deficiencies in the *Work* for which the *Contractor* is proven responsible.

As used herein "substantial defects or deficiencies" means those defects or deficiencies in the *Work* which affect the *Work* to such an extent or in such a manner that significant part or the whole of the *Work* is unfit for the purpose intended by the *Contract Documents*.

. . .

Whereas the CCDC form may prescribe a relatively short warranty period of one year, the longer six-year period from substantial performance for substantial defects or deficiencies exposes the contractor to claims for a much longer period. The six-year limitation period from substantial performance pursuant to the CCDC standard form contract may seem onerous to many contractors. However, it is nevertheless more advantageous to the contractor than the common law in Ontario that could impose a limitation period, in simple contract, of six years from when the damage is first discovered or ought reasonably to have been discovered, where the contract involves a duty of care. Owners who accept a contractual right of lesser duration than the open-ended common-law principle would provide typically rationalize that some time limit is reasonable and that, in any event, it is not an

unreasonable risk to expect that substantial defects or deficiencies will likely turn up within such six-year period subsequent to
substantial performance.

CHAPTER TWENTY-SIX

BONDS AND INTERNATIONAL PERFORMANCE GUARANTEES

Bonds are commonly used on construction projects and in many equipment-supply contracts. A basic understanding of bonding is important to the engineer. The most common forms of bonds are the bid bond, the performance bond, and the labour and material-payment bond.

Bonds are contracts, the bonding company (the "surety") agreeing to guarantee the performance of a specified contractual obligation. In common bonding terminology, the "principal" is the party whose performance of the contractual obligation is guaranteed by the bonding company, or surety. The principal is normally a contractor or subcontractor. The term "obligee" is used to describe the party for whose benefit the bond is provided. The obligee is often the owner.

It is extremely important to appreciate that a bond is not insurance. If the bonding company is required to incur expense in performing its obligations, it will seek to recover its expenses from the principal.

Often the principal (contractor) is a corporation. Bonding companies often require shareholders or principal officers of a corporation that is named as principal in a bond to indemnify the surety against any expense incurred by the surety. The bonding company may then look to both the corporation, as principal, and to the individuals who provide the indemnities to recover any payments the bonding company may be required to make as it fulfils its obligations to the obligee.

Bonds are essentially contracts among three parties, the surety, the principal, and the obligee. In effect, the surety indemnifies the obligee against loss arising from the principal's failure to perform contractual obligations.

BID BOND

The bid bond is used in tendering. If the principal's tender or bid is accepted, the principal is required to enter into a formal contract with the obligee. If the principal fails to enter into the formal contract, he or she must pay to the obligee the difference in money between the amount of the bid and the amount the obligee must pay another party for the work, up to the face amount of the bid bond. Should the principal fail to pay, the surety has to pay it. However, the surety will then charge the total costs involved back to the principal.

PERFORMANCE BOND

A performance bond indemnifies the obligee if the principal doesn't perform his or her contractual obligations. The indemnity, however, is limited to the amount specified in the performance bond. The performance bond is normally provided when the formal contract is executed. The performance bond identifies the contract that the principal must fulfil. Sometimes performance bonds are for all of the contract price; or they can be for a limited amount. It is very important to review and understand and be satisfied with the performance bond wording. The surety's obligations are limited to that wording and the contractual obligations to which the performance bond applies.

The amount of the performance bond relates to the extra cost above the original contract price that may be required to complete the contract if the bonding company is obligated to do so. Accordingly, upon default by the contractor and new arrangements being made to have another contractor finish the work, the owner will still be obligated to pay an amount equal to the balance of the original contract price. The surety's obligation to provide funds for construction applies to amounts required over and above the original contract price.

The most straightforward circumstances to which the performance bond typically responds will be when the contractor is adjudged bankrupt or a receiver is appointed on account of the contractor's insolvency and the receiver acknowledges that the contractor is unable to continue to perform. In those circumstances the surety usually steps in and makes arrangements for the contract to be completed, pursuant to the provisions of the performance bond. However, short of those circumstances, if an owner claims that the contractor is defaulting in performance and the contractor disputes the owner's claim, the resolution of the matter is likely to be prolonged. The reason is straightforward. The bonding company needs to be satisfied that there has been a default in performance before it will step in. The bonding company may reasonably take the position that the issue of performance first be resolved by a court or arbitration process. The contractor will usually strongly object to the allegations of default, particularly as the contractor's principals will likely be obligated to indemnify the surety.

At the time of tendering, it is common practice to require bidders to provide agreements to bond, or consent letters that confirm that the surety will provide performance or other bonds pursuant to the tender instructions.

LABOUR AND MATERIAL-PAYMENT BOND

The purpose of a labour and material-payment bond is to guarantee that all claimants will be paid for labour and materials provided to the principal for the project described in the bond. In the traditional contractual arrangement between an owner and a general contractor, the potential claimants are usually subcontractors and material suppliers who have contracts directly with the principal (contractor). Because there is usually no contract between the bonding company and such claimants, the owner should be described, in the labour and material-payment bond, as the obligee who acts as trustee for the benefit of such claimants.

The form of bond chosen is most important. Bonds should be reviewed by legal counsel where necessary.

LETTERS OF CREDIT

The use of letters of credit on construction contracts and on equipment-supply contracts is becoming more common, particularly when a job is to be performed outside Canada. For example, someone might ask for a letter of credit instead of a performance bond. Letters of credit are normally provided by chartered banks or trust companies. The letters provide an assurance that funds are available if, for example, a contractor defaults in performance. Depending upon their terms, letters of credit can amount to "blank cheques" for the person in whose favour the letter is written and legal advice should be obtained where unusual terms are encountered. Often the terms of the letter of credit will provide that the bank or trust company must pay the holder of the letter if he or she provides a certificate or affidavit saying a contractual default has occurred. Being required to provide a letter of credit can significantly complicate matters for a contractor as it normally impacts directly on the contractor's line of credit with the bank or trust company.

INTERNATIONAL PERFORMANCE GUARANTEES

In many foreign jurisdictions, particularly in the Middle East, Africa, and Southeast Asia, demand letters of credit are in fact referred to as "performance bonds" or "performance guarantees" and are typically required on engineering projects. Given the alleged and potential abuse of demand letters of credit required by owners on international construction projects, it is extremely important to exercise caution prior to agreeing to provide such security. In response to concerns about "bad calls" on performance guarantees, alternative forms of security have been considered. One such alternative would be a conditional letter of credit, payable subject to an arbitrator's or referee's decision. Naturally, owners far prefer unconditional letters of credit.

The need for making use of a provisional decision maker on international construction project disputes was one of many issues that surfaced in the development of the International Bar Association's Illustrative Forms of Performance Guarantee and Counter Guarantee Security for International Construction Projects (the "IBA's performance guarantee forms"). Recognizing the need for an independent interim decision maker, the IBA's performance guarantee makes use of the Pre-arbitral Referee Procedure of the International Chamber of Commerce in Paris (the "ICC").

The IBA's performance guarantee forms, although sometimes referred to as "performance bonds," are not at all "performance bonds" by North American standards. Nor do they provide any guarantee of specific performance by common-law standards. Rather, the IBA's performance guarantee forms are a means of implementing conditional letter of credit security arrangements, intended to increase the likelihood of a more balanced, "even-handed" approach to project security than demand letters of credit that create the potential for arbitrary calls by owners; arbitrary calls which aggrieved contractors report to have experienced on international projects.

The concept contemplated by the IBA performance guarantee forms is reasonably straightforward. The contractor's bank provides a performance guarantee that entitles the employer (or owner) to demand payment by delivering to the contractor's bank a demand stating that the contractor is in material breach of contract, identifying the nature of the breach, and specifying the amount payable to the owner as damages for the contractor's material breach. Upon such demand, three alternatives then arise that relate to payment to the owner:

(1) If the contractor doesn't object, the owner is paid.

(2) If the contractor objects, the owner is paid as soon as the owner provides a counter-guarantee. The counter-guarantee entitles the contractor to be repaid if and when an ICC pre-arbitral referee decides in the contractor's favour.

(3) If the contractor objects and a counter-guarantee is not provided by the owner, the owner is paid if and when an ICC pre-arbitral referee decides in the owner's favour.

Payments under the performance guarantee or under the counter-guarantee are made only on presentation of specific documents, in agreed form, to the appropriate bank.

ADR, through the ICC Pre-arbitral Referee Procedure, was chosen because of the ICC's stated intent to provide a mechanism for swift provisional decisions, an obviously important feature where disputes require resolution on a timely basis so as to minimize, or at least reduce, the impact of the dispute on the scheduling of the project.

The independent nature of the ICC pre-arbitral referee should provide some additional comfort to international parties, particularly in civil law jurisdictions where project participants often resist attempts to implement the traditional contractual common-law approach of designating the project "consultant" as the interim decision maker. The provisional decision of the pre-arbitral referee does not preclude further proceedings on the issue between the parties.

The IBA's performance guarantee forms, as illustrative forms, provide a guide for implementing this approach as an alternative to demand letters of credit. Not surprisingly, owners typically resist the concept of anything less than a demand letter of credit and, accordingly, the "marketability" of the IBA performance guarantee forms or of any "conditional" letters of credit (that require a referee's or arbitrator's decision before payment) will turn on the realities of the marketplace and the negotiating skills of the project participants.

CHAPTER TWENTY-SEVEN

SUBSURFACE ISSUES

Perhaps not surprisingly, given the emphasis in the previous chapter on risks in construction, Canadian courts have often dealt with cases involving issues relating to subsurface conditions. As difficult a dilemma as subsurface risks may present, Canadian cases confirm the strict approach the courts take to enforcing clear contractual obligations, even those undertaken in circumstances when the bidders may well have had far less than optimum conditions upon which to investigate, assess, and price the risks. However, where Canadian courts have decided in favour of the contractor claiming additional compensation on account of variations in subsurface ground conditions, a key ingredient has been substantiating that the project owner and/or engineer failed to disclose important information to the bidders.

The failure of an owner to disclose important information to bidders is now recognized as a basis for a contractor to claim additional compensation, subject of course to agreed contract terms. This chapter expands the previous review of the general basis upon which parties contract with respect to subsurface conditions and also reviews, by illustrative case summaries, the principles that courts have applied in deciding who assumed what risks as far as ground conditions are concerned.

In spite of the CCDC provisions relating to subsurface conditions as described previously, sometimes project owners contractually shift the subsurface risks to the contractor, to a very substantial degree. Owners who do so risk substantially inflated bids by contractors seeking to protect themselves from unknown subsurface conditions. Contractors face the difficulties of building into their prices an amount, often referred to as a "contingency," to attempt to cover the risk. In doing so, they risk putting in too high a price.

The following is an example of a ground conditions clause that is very advantageous to the owner:

> The Contractor acknowledges that it has visited the site of the work prior to agreeing to enter into the contract and has satisfied itself by its own examination or otherwise as to the local conditions to be met during its performance of the work, including, without limitation, an examination of subsurface materials and conditions. The contractor agrees that under no circumstances whatsoever will it make a claim for reimbursement on account of changes in subsurface conditions which it may claim to encounter during its performance of the work.
>
> No reports of subsurface investigations made available by the Owner or its agents or representatives shall limit or restrict the Contractor's obligation pursuant to paragraph 28.1 aforesaid to perform the work notwithstanding the nature of subsurface materials and conditions or the extent to which such subsurface materials or conditions may vary from that indicated by any such reports. The Contractor acknowledges that the Owner and its agents and representatives do not accept responsibility for the accuracy of any information contained in any such report relating to subsurface investigations. The Contractor further acknowledges its obligation to itself to have conducted or caused to be conducted all such investigations as may be necessary to inform the Contractor of subsurface conditions which may be encountered during its performance of the work. No change in subsurface conditions shall entitle the Contractor to additional reimbursement on account thereof.

By contrast, the following is an illustration of a more reasonable clause, from the contractor's perspective:

CHANGES IN SOIL CONDITIONS

If the Contractor incurs or sustains any extra expense or any loss or damage that is directly attributable to:

a substantial difference between the information relating to soil conditions at the work site that is contained in the plans and specifications or other documents supplied to the Contractor for its use in preparing its tender or a reasonable assumption of fact based thereon made by the Contractor, and the actual soil conditions encountered by the Contractor at the work site during the performance of the contract, or

. . .

it shall, within ten days of the date that an event described in GC . . . occurred, give the Engineer written notice of the event and of its intention to claim for that extra expense or that loss or damage.

If the Engineer determines that a claim referred to in GC . . . is justified, the Owner may make an extra payment to the Contractor in an amount that is calculated in accordance with GC . . . .

Prior subsurface investigation by qualified experts is advisable. However, as the experienced project participant is well aware, all subsurface conditions may be very difficult to ascertain even where such an investigation has been carried out. That being the case, it is difficult to argue with the reasonableness of recommending that a contractor be entitled to additional compensation where subsurface conditions differ materially from those indicated by information or reports included in the contract documents, or from a reasonable assumption of probable conditions based thereon.

CONTRACTOR RESPONSIBLE FOR QUANTITY ESTIMATES

The following cases illustrate some circumstances where contractors have been denied claims for additional compensation relating to subsurface ground conditions:

1. A 1987 decision of the Federal Court of Canada (Trial Division) involved specifications containing a warning that the excavated material might prove to be unsuitable and result in substantially more borrowed fill being required than estimated. The case was *Dexter Construction Company* v. *Her Majesty The Queen In Right Of Canada As Represented By Her Minister Of Public Works*.[1] The tender documents for the road reconstruction referred to two additional documents that were not included in the tender package, a soil report and a mass haul diagram.

---

[1] 29 C.L.R. 124

These showed sources, quantities, and disposition of materials, which were erroneous. The evidence showed the contractor hadn't reviewed the soil report nor the mass haul diagram. Instead the contractor made its own observations on site, studied the other tender documents, and assumed that a substantial quantity of the excavated material could be used as fill. That was not the case and the contractor claimed damages for negligent misrepresentation, on account of changed soil conditions, or *quantum meruit* on the basis that the contractor was entitled to repudiate the contract in view of the significant changes in the nature of the work. The Court held, in spite of the errors, that the defendant had effectively discharged its duty to warn that substantial quantities of additional borrow might be required. The contractor hadn't relied on the erroneous soil report and the mass haul diagram in any event. Accordingly, the contractor's claim for additional compensation on the changed soil conditions clause failed. On the *quantum meruit* claim, even though parts of the project had been subject to significant changes because of unsuitable excavated material, the Court decided that overall the project quantities were reasonably close to the estimates (except for the quantities to which the specific warning related) and, accordingly, the contractor was not entitled to repudiate and claim on the *quantum meruit* basis.

2.   In its 1982 decision in *Carman Construction Ltd.* v. *Canadian Pacific Railway Co. et al.*[2] the Supreme Court of Canada dismissed the appeal of the contractor who claimed additional compensation for rock excavation pursuant to an upset price contract. The contractor was contractually obliged to calculate the volume of rock to be removed in widening an embankment adjacent to a railway line.

The contract contained the following clause:

3.1 It is hereby declared and agreed by the contractor that this Agreement has been entered into by him on his own knowledge respecting the nature and conformation of the ground upon which the work is to be done, the location, character, quality and quantities of the material to be removed, the character of the equipment and facilities needed, the general and local conditions and all other matters which can in any way affect

---

[2]   136 D.L.R. 193

the work under this Agreement, and the contractor does not rely upon any information given or statement made to him in relation to the work by the Company.

Prior to bidding, the contractor was told that no soils report or cross-sections were available. However, the facts revealed that an employee of CPR had volunteered a volume estimate to the contractor of 7000–7500 cubic yards of rock based on a 2500 foot excavation length. The contract indicated the excavation length to be 2950 feet. Unfortunately for the contractor, it prepared its bid using the 7500 cubic yard estimate and 2500 foot excavation length. The contractor argued that the tender package did not disclose the volume of rock to be removed and that there was not enough time in the circumstances to have a consulting firm undertake an investigation to accurately determine the quantity.

The evidence disclosed that the contractor didn't know the name of the employee who provided the volume estimate, what the person's position was, or what authority he had. However, the judge was satisfied that the information had been provided by an employee of the owner.

The Supreme Court of Canada confirmed that the contractor was aware of its contractual obligations with respect to the quantities of material to be removed. The Court emphasized that the written agreement governed and that the employee's misrepresentation of the measurements did not amount to a collateral agreement given the express contract language. The Court also held that the disclaimer clause precluded a duty of care on the part of CPR and accordingly it wasn't liable for negligent misrepresentation.

3. *Diamond Willow Ranch Ltd.* v. *Oliver (Village) et al.*[3] a 1988 decision of British Columbia's Supreme Court, is another case that illustrates the contractor's inability to recover where the contract required the contractor to study and examine the site and the conditions. The facts are not at all impressive, as far as the contractor's investigation of conditions was concerned. In bidding on three earth-filled dams in connection with a sewage effluent project, the contractor failed to familiarize itself with site topography, local weather conditions, or soil information. Having failed to do so, it's perhaps not surprising that

[3] 31 C.L.R. 287

difficulties were encountered. Even though conditions proved difficult the Court concluded that they were not materially different from those specified and the Court refused the contractor's claim.

DAMAGE TO OTHER EXISTING FACILITIES

A significant risk on many construction projects can be the potential damage to existing buildings and structures located in the general area of the construction. Two examples are settlement caused by dewatering operations and loss of lateral support.

Liability for negligence and/or nuisance for damages resulting from dewatering operations is illustrated by the Ontario Court of Appeal's 1977 decision in *Pugliese et al.* v. *National Capital Commission et al.*[4] In that case, the owners of 101 residential properties claimed that the ground water table below their properties was substantially lowered by the defendants during the construction of a collector sewer on lands located near their homes. During construction subsurface water had been pumped from deep drainage wells. The plaintiffs claimed that the dewatering operations caused the consolidation of the underlying soil, which resulted in differential settlement and serious damage to their homes. The court held that the defendants had a duty to use reasonable care in extracting the ground water. However, the court also held that taking all reasonable care would not necessarily be enough as far as nuisance was concerned. A landowner would have a claim in nuisance if the damage to its land was beyond that which it could reasonably be expected to tolerate.

Gallant et al. v. *F.W. Woolworth Co. Limited et al.*[5] is a 1970 decision of the Saskatchewan Court of Appeal, which illustrates how an owner can be liable for damage to an existing building caused by a loss of lateral support. The defendant owner hired a contractor to construct a building separated from the plaintiff's building by a laneway. In preparing to pour the footings, the contractor made an excavation into the laneway that was not properly compacted on backfilling as required by the contract. Shortly after the roof of the plaintiff's building began to leak and the walls cracked. The plaintiff succeeded in its claim against the owner and contractor for damages as a result of loss of lateral

---

[4]   79 D.L.R. (3d) 592
[5]   [1971] W.W.R. 336

support. The Court of Appeal stated that when lateral support to land is removed, it is immaterial whether the act causing the damage was negligent.

DUTY TO DISCLOSE CORRECT INFORMATION

Courts have found in the contractor's favour, where incorrect information has been provided by the owner or engineer. An example is the 1984 decision of the Ontario Supreme Court in *Cardinal Construction Ltd.* v. *Corporation Of The City Of Brockville et al.*[6] The contractor responded to a tender call. The drawings included in the tender package had described certain "underground Bell cables," understood by all parties to be a two-inch flexible plastic covered cable. However, what was actually encountered was a concrete encased duct structure that required the relocation of a water main. The result was a substantial delay and additional costs to the contractor.

The Court held that the contractor was justified in relying on the information originally supplied by the owner's engineer even though the contract required the contractor to investigate underground utilities and the character of the work prior to tendering or assume the risk of unanticipated or onerous conditions. The court held that the owner had breached its duty to present accurate information or to warn bidders to verify information and that the owner's negligent misstatements constituted torts.

Another decision involving a failure to disclose was the 1989 decision of the British Columbia Supreme Court in *Hartnett* v. *Wailea Construction.*[7] The case involved the construction of a house on a former landfill site. As the lot had been used as a landfill site, the municipal by-laws required a soil report for the purpose of subdivision review. The vendor commissioned a soil report; however, the lot purchaser did not become aware of the soil report until after the foundation was in place.

The court imposed liability on the vendor because of its fraudulent misrepresentation. The municipality was liable for breach of its duty to disclose soils information in its possession.

As previously referenced, in the 1988 decision of the Supreme Court of British Columbia in *B.G. Checo International Ltd.* v. *British Columbia (British Columbia Hydro and Power*

[6]   4 C.L.R. 149
[7]   33 C.L.R. 244

Authority),[8] B.C. Hydro's failure to disclose important information in the tender documents has been characterized as a form of "tender by ambush." The contract involved the construction of two new transmission lines in British Columbia, primarily involving the erection of towers. B.C. Hydro failed to disclose in its tender document that logs and debris would remain on the right-of-way causing delay and additional cost. B.C. Hydro argued that the existence of the logs on the right-of-way would be obvious to a tenderer viewing the site. The specifications provided to B.C. Hydro's Construction Department clearly stated that "logs and old loggings slash will be on the right-of-way in some areas," but this statement was omitted in the tender package distributed to bidders. The Court was satisfied that B.C. Hydro's omission constituted a fraudulent misrepresentation.

METHODS DICTATED BY CONTRACT DOCUMENTS

In *Warden Construction Co. Ltd.* v. *Corporation of The Town of Grimsby,*[9] a 1983 decision of the Ontario Court of Appeal, the contractor avoided subsurface risks in carrying out the work because the work method stipulated in the plans and specifications was impossible of performance. The case involved the construction of a water main. The installation was specified to be by the boring and jacking method. The contract, however, contained the following provision advantageous to the owner:

> "Contractor's Investigations: The contractor declares and represents that in tendering for the work, and in entering into this Contract, he has either investigated for himself the character of the work to be done and all local conditions, including the location of any utility which can be determined from the records or other information available at the offices of any person, partnership, corporation, including a municipal corporation and any board or commission thereof having jurisdiction or control over the utility, that might affect his Tender or his acceptance of the work, or that, not having so investigated, he is willing to assume and does hereby assume, all risk of conditions now existing or arising in the course of the work which might or could make the work, or any items

---

[8] Supra, at page 168
[9] 2 C.L.R. 69

thereof more expensive in character, or more onerous to fulfil, than was contemplated or known when the Tender was made or the Contract signed. The Contractor also declares that in tendering for the work and in entering into this Contract he did not and does not rely upon information furnished by the owner or any of its servants or agents respecting the nature or conformation of the ground at the site of the work, or the general and local performance of the work, under the Contract."

As it was impossible for the contractor to install the water main by the specific method of boring and jacking, installation was carried out by tunnelling and hand mining. This involved greater expense. The Court accepted the argument that as the performance of the contract in the manner specified in the plans and specifications was impossible, the contract provision did not apply. Accordingly, the contractor succeeded in its claim for extra costs. The case points to the dangers to owners and their engineers of specifying actual construction methods. Typically, construction contracts provide that the contractor controls the selection of construction methods.

The material in this Chapter 27 is based on a paper presented by D.L. Marston to the International Construction Projects Committee of the International Bar Association (the "IBA") in Strasbourg, France in 1989 and is included in this text with the kind consent of the IBA.

CHAPTER TWENTY-EIGHT

ARBITRATION AND ADR

A lawsuit is not always the best way to resolve a dispute between contracting parties, particularly disputes of a "technical" engineering nature. The dispute may also be resolved by a somewhat less formal procedure called "arbitration." Arbitration has developed to provide an alternative that is intended to be less costly, less protracted, and less public than litigation. Arbitration is now a common approach to dispute resolution, in Canada and abroad. It is also part of the leading-edge focus on approaches to alternative dispute resolution ("ADR") on engineering projects. The engineer is often involved in arbitrations. The engineer may be party to the dispute, an expert witness, or the arbitrator. An engineer who acts as an arbitrator is expected to act entirely impartially and independently of the parties to the dispute.

Arbitration of disputes of a technical (engineering) nature is very sensible, particularly as it provides an opportunity to bring the dispute before an arbitrator who is already familiar, generally at least, with the technical nature of the subject matter in dispute.

An arbitration provision is usually included in engineering and construction contracts. The wording of the arbitration provision may indicate that arbitration is mandatory — that is, the parties must submit disputes to arbitration; they may not proceed to resolve the dispute by way of a lawsuit. An example of a mandatory arbitration provision is as follows:

> All matters in difference between the parties hereto in relation to this agreement shall be referred to arbitration.

But not all arbitration provisions indicate mandatory arbitration; the clause might provide that disputes will be submitted to arbitration if both parties agree to so arbitrate upon the request of one of the parties. Such a clause is simply notice that arbitration is available as a means of resolving a dispute.

APPOINTMENT OF ARBITRATOR

Some contracts describe the manner in which an arbitrator is to be appointed and detail the general procedure that will govern the arbitration. For example, an arbitration clause may provide that each party to the dispute shall appoint a representative arbitrator, and that the two representative arbitrators shall then appoint a chair, thereby creating a three-person arbitration board. The contract may also provide that, if the two representative arbitrators cannot agree upon the selection of the chair, the chair will be appointed by a court.

THE ARBITRATION ACT

Engineering and construction contracts usually provide that arbitration will be governed by a provincial arbitration statute, typically the statute of the province whose law governs the interpretation of the contract. The Arbitration Act, 1991 of Ontario[1] is but one example of such a statute. The Act deals with the appointment of an arbitrator or arbitrators; and it sets out a structure or set of rules to govern the conduct of an arbitration unless the parties agree that some other structure or rules will apply. For example, the Arbitration Act sets out rules about such matters as the time and place of arbitration, the commencement of arbitrations, and the exchange of pleadings. By settling a framework of rules, the Arbitration Act helps to mitigate the costs, delays, "gamesmanship," and uncertainty that could arise if that was left to subsequent negotiation. However, the parties to an arbitration can always agree to vary the rules set forth in the Arbitration Act.

The Arbitration Act limits the circumstances in which the courts may interfere in the arbitration process or review the arbitration award, compared to the previous Arbitration Act. Prior to January 1, 1992, arbitration awards in Ontario could be impeached by an application for judicial review or, if the arbitration submission so provided, on appeal. These judicial review and appeal procedures gave the unsuccessful party the opportunity to prolong the dispute and delay having to pay the successful party. The Arbitration Act limits the range of circumstances in which the arbitration award can be impeached. With respect to appeals, the

[1]   S.O. 1991, c. 17

Arbitration Act provides that, in the absence of a contrary provision in the arbitration agreement, the party may appeal on a question of law only with leave of the court. The Arbitration Act allows parties to expand, narrow, or exclude that limited right of appeal. With respect to applications for judicial review, the Arbitration Act now limits the grounds on which an award may be set aside to matters clearly related to the conduct of the arbitration rather than the merits of the award. Section 46(1) provides:

> 46(1) On a party's application, the court may set aside an award on any of the following grounds:
> 1. A party entered into the arbitration agreement while under a legal incapacity.
> 2. The arbitration agreement is invalid or has ceased to exist.
> 3. The award deals with a dispute that the arbitration agreement does not cover or contains a decision on a matter that is beyond the scope of the agreement.
> 4. The composition of the tribunal was not in accordance with the arbitration agreement or, if the agreement did not deal with that matter, was not in accordance with this Act.
> 5. The subject-matter of the dispute is not capable of being the subject of arbitration under Ontario law.
> 6. The applicant was not treated equally and fairly, was not given an opportunity to present a case or to respond to another party's case, or was not given proper notice of the arbitration or of the appointment of an arbitrator.
> 7. The procedures followed in the arbitration did not comply with this Act.
> 8. An arbitrator has committed a corrupt or fraudulent act or there is a reasonable apprehension of bias.
> 9. The award was obtained by fraud.

Accordingly, the Arbitration Act should help to reduce the hazard that an arbitration hearing will merely be treated by one of the parties as a prologue to further and expensive proceedings in court. The Arbitration Act provides a better basis to have the dispute resolved — and finally resolved — by way of arbitration.

The Arbitration Act enhances the integrity of the arbitration process by codifying the common law rule that an arbitrator shall be independent of the parties and shall act impartially. Arbitrators cannot consider themselves the advocates of the parties who appointed them.

The Arbitration Act expands the powers that arbitrators may exercise. For instance, prior to the Arbitration Act, there was some uncertainty about the jurisdiction of arbitrators to award interest. The Arbitration Act provides that arbitrators now have the same jurisdiction to award prejudgment and postjudgment interest as courts have under the Courts of Justice Act. The Arbitration Act also expands the scope of remedies that arbitrators may grant. In particular, the Arbitration Act expressly confers on arbitrators the power to grant equitable remedies such as injunctions and specific performance.

NEW APPROACHES TO RESPOND TO DISPUTE RESOLUTION DIFFICULTIES ON CONSTRUCTION PROJECTS

ADR has sometimes been referred to as a revolution — the "alternative dispute resolution" revolution. Inspired by the time-consuming, disruptive impact of enormously costly construction litigation (and, at times, arbitration), the quest for an alternative approach to dispute resolution has become a central focus of interest among sophisticated construction industry participants.

Partnering on Infrastructure and Construction Projects

"Partnering" is currently a commonly used term and concept that is intended to respond to the need for improved attitudes amongst construction industry participants to the importance of teamwork on projects. Partnering is a concept that is aimed at promoting cooperation among all project participants. Whereas partnering is not technically an integral part of ADR programs, it is certainly reflective of the objectives of ADR. Partnering is an effort to educate all participants on a project on the mutual benefits that can be obtained by working cooperatively as a team toward common goals through the establishment of a healthy, cooperative attitude at the outset. This cooperative attitude should assist in resolving project disputes at as early a stage as possible.

In order to implement partnering, a partnering "workshop" is conducted subsequent to the award of the contract. The partnering workshop provides a forum in which participants focus on the importance of good communications, team spirit, and mutual project goals. The partnering document is not intended to replace the project contract but to outline, in a very general way, important cooperative goals. For partnering to work, the commitment of senior management is very important. That commitment obviously sends a significant message to other representatives of the respective parties involved in the project.

Project Neutral

Another approach that has been implemented, primarily in the U.S. to date, is consistent with involving Dispute Review Boards on an ongoing basis throughout a project. This approach involves the appointment of a project neutral, typically an independent professional experienced in the construction industry, to stay abreast of developments on the project with a view to offering advice and decisions on an unbiased basis.

The project neutral may be retained to assist in the resolution of disputes between parties. Conceivably the project neutral could be authorized to make binding decisions with respect to certain issues and disputes.

The use of independent consultants on projects, for example by owners on design-build projects, is not a new concept. However, further consideration to expanding the application of this concept could be given if parties are genuinely serious about achieving a basis upon which disputes are to be finally resolved at an early stage in the process — particularly disputes of a technical nature such as those relating to the interpretation of technical specifications, scheduling extensions, the value of change orders, and technical deficiencies in the work.

Involvement of a project neutral or other independent consultant or team of independent consultants has sometimes been referred to as project management overview. A primary reason for reluctance to implement such approaches in Canada, it seems, is the extra cost involved in doing so. Presumably, with an understanding of the advisability of implementing preventive project approaches and ADR techniques, the economic benefit of involving unbiased independents on appropriate construction projects will achieve greater acceptance.

Mediation

Participants in the construction industry are looking more and more to mediation (or conciliation) as an important step in achieving a negotiated settlement of disputes.

To be successful, parties to mediation need to perceive some advantage to resolving the dispute through a negotiation process involving an impartial mediator who does not act as arbitrator or judge but is there to provide guidance to the parties and to facilitate the settlement process by acting as a "go-between" in communicating proposed settlement positions.

A fundamental difference between litigation and arbitration and other forms of dispute resolution is that it is only litigation and arbitration that result in an adjudication that is binding. In mediation, it is up to the parties to work out their differences.

CHAPTER TWENTY-NINE

ADR ON INTERNATIONAL PROJECTS

As already emphasized, more and more attention is currently being focused on the evolution of special techniques to address the reduction and resolution of disputes, both on Canadian and international projects, be they "BOT," infrastructure, or traditional construction projects. Examples of techniques aimed at reducing disputes include "partnering" concepts, dispute review boards, pre-arbitral referees, and contractual mediation provisions.

The focus of attention on ADR was primarily generated from the U.S. However, it is now a topic of frequent discussion among industry participants in many countries around the world. Not surprisingly, acceptance and application of ADR is evolving through varying stages of development in different countries and from project to project. A central theme of North American construction law conferences, ADR has also been a central theme of focus for a number of years at international construction law conferences in many parts of the world. Growing international awareness of the importance of ADR in construction continues to generate even more attention. That world-wide attention and growing awareness are resulting in the evolution of various ADR approaches adapted to attempt to avoid, or at least to minimize, the disruptive and costly impact of traditional methods of dispute resolution.

The emphasis in this chapter will be to consider the trends in ADR on infrastructure projects by reviewing, briefly, the nature of the problem giving rise to the need for ADR, and to consider the approaches that have been taken to ADR on large international infrastructure projects.

An awareness of the approach on internationally tendered infrastructure projects is becoming increasingly important to Canadians, particularly as more and more Canadian construction industry participants look to other parts of the world for major project opportunities. Hopefully this review of international trends will provide information to Canadian participants that will assist them in sorting out appropriate approaches to be taken to ADR, both in Canada and abroad.

A primary focus in this chapter will be to consider the trend towards recognizing decisions of project neutrals and dispute review boards as final and binding, rather than as provisional or interim decisions on disputes arising during the construction process. Provisional or interim decisions leave it open to the parties, if one of them is not satisfied with the provisional decision, to subsequently dispute the decision and submit it to a further arbitration or judicial process for final determination.

RECOGNITION OF THE PROBLEM

At a meeting of the American Bar Association's Forum on the Construction Industry held in Charleston, South Carolina in April, 1994, the nature of the problem was identified in the following introductory comments by one of the program speakers:

> The construction industry in the '90s is undergoing a profound change. In their joint invitation letter to participants in the 1993 Conference, the Chairs of the ABA Forum on the Construction Industry and of this Conference called attention to the construction industry's "growing discontent with the status quo — too many disputes; too much time at too much cost." This discontent has in recent years propelled the industry into a new era of research, experimentation and emphasis into restructuring cooperative project relationships and developing techniques for the early resolution of problems.

Concerns in Canada are consistent with those expressed in the U.S., although the move to deal with ADR issues has trailed behind the U.S. initiatives. Nevertheless, the Canadian construction industry, having suffered through similar problems to those experienced in the U.S., is responding to the need for ADR. Evidence of this response is reflected in recently proposed amendments to the provisions of the Canadian Construction Documents Committee's contract form and also in initiatives of the Ontario government resulting in the establishment of the Attorney General's Advisory Group on Alternative Resolution of Construction Disputes. Ontario's former Attorney General, Howard Hampton, acknowledged in 1992 that:

> Over the years people in many segments of the construction industry, and some of its legal advisers, have voiced dissatisfaction with reliance on the court process as a means of resolving construction-related disputes. They have complained that: the court process is too slow; the judges lack expertise in the field; reliance on an adversarial process to resolve a dispute can impede the ongoing working relationships on a project.
>
> It makes sense for the Ministry to encourage and assist private interests to voluntarily remove from the court system types of disputes that are better resolved by other means.

As experienced construction industry participants are well aware, there are a variety of factors that influence the potential for disputes on infrastructure projects, not the least of which are a typically large number of participants involved in the project, the contractual interrelationships among those participants, and the propensity for "finger pointing" among project participants when problems arise, not to mention the high stakes that are typically involved on major infrastructure projects. These factors have been reflected in a discussion paper prepared by the Ministry of the Attorney General Advisory Group on Alternative Resolution of Construction Disputes. This paper identifies, among leading causes of disputes, salient factors including: aggressive contracting approaches that attempt to unreasonably shift risks between project participants; ambiguities in contract documents; insufficient financing; poor communication; inadequate contractor management and coordination; inability to deal promptly with changes in unexpected conditions; litigious "mind-sets"; a lack of willingness on the part of some contract administrators to make decisions to resolve problems at the outset; and last, but certainly not least, a lack of team spirit amongst project participants. The Advisory Group's identification of significant causes certainly rings true. It is obviously advisable for the Canadian construction industry to implement an ADR system in order to alleviate the significant problems that have developed.

The realities of aggressive contracting approaches on Canadian projects are well known to industry participants. However, it is important to bear in mind, in recognizing the need for ADR, the nature and extent of aggressive adversarial approaches to contracting that have been reflected in contracts on Canadian projects. Paranoid owners, determined to do whatever possible to avoid or minimize claims from contractors, have gone so far as to include express contractual wording to preclude contractors from making claims for additional reimbursement even in circumstances where the cause of the contractor's complaint has been the owner's delay in providing approvals or in fact the owner's interference with the work of the contractor. Another example of aggressive contracting approaches by owners is to include provisions that unrealistically shift the risk of variations in subsurface conditions to contractors. Some contractors, on the other hand, have responded by marshalling aggressive approaches to submitting claims for extras on projects. Canadian projects have seen the use of contractors' formalized "claims teams" organized on-site, with legal counsel forming part of these teams. In addition, the realities of the fast-track approach to major construction projects provide fertile ground for disputes over issues relating to design scope. These are some examples of aggressive contracting realities that have elevated the potential for construction litigation, particularly in the United States and in Canada.

THE CONSULTANT'S CONFLICT

A further observation that gives rise to a weakness, in theory at least, in the traditional approach to the resolution of disputes in Canada is that the resolution of on-site disputes is contractually left to the decision of the "Consultant," the individual (engineer or architect) designated as the contractor administrator in the construction contract between the owner and the contractor. Typically, the consultant is also the designer of the project. When disputes over the quality of construction arise, fingers are usually pointed by contractors and subcontractors to alleged design deficiencies. This often results in the consultant, as the decision maker of first instance on the dispute, being faced with making a decision that relates to construction and also to design issues. Obviously, the consultant who is responsible for the design may then well face a conflict issue. This potential for conflict has often been identified as a weakness in the contracting approach adopted widely in the U.S. and in Canada, and also reflected in international forms such as the FIDIC form. Those who advocate this approach emphasize that, by and large, the system nevertheless works reasonably well; that consultants have in fact handled the initial decision-making process well enough (such initial decisions typically being subject to subsequent arbitration or litigation in any event); and, accordingly, that the approach does work. This explanation is not always a satisfactory one, particularly to construction counsel in civil law jurisdictions where the importance of an independent decision maker often receives greater attention and emphasis.

Recognizing the need for a fresh approach to dispute resolution on Canadian contracts, the Canadian Construction Documents Committee (the "CCDC"), has now included in its revisions to CCDC, a new dispute resolution approach. The approach provides that claims relating to the performance of the work or the interpretation of the contract documents are to be initially determined by the consultant in writing and, unless a written dispute notice is provided within fifteen working days, the parties are conclusively deemed to have accepted the consultant's decision. However, where a notice of dispute is provided, the new CCDC contract form obligates the parties to proceed to mediate such dispute in accordance with the provisions of the CCDC contract. Should mediation not prove successful in finally resolving the dispute, one of the parties to the dispute would then be entitled to require binding arbitration. If neither party chooses to proceed with binding arbitration, litigation is open to the parties. In implementing CCDC contract forms that include such provisions, it will be extremely important for the contracting parties to pay close attention to the administration of the contract in accordance with its terms. Close administration of a construction contract is always advisable for a host of reasons. However, it will be particularly important in order that a party avoids being deemed to have accepted a consultant's decision with which it disagrees and so loses an opportunity to require mediation or to require arbitration if desired.

Note that criticism in North America is most often directed at the outcome of the litigation process. Elsewhere in the world, where arbitration of construction disputes is the norm, some similar criticisms have also been directed at the arbitration process. However, it makes eminently good sense to bring technical construction disputes before arbitrators who are familiar with the industry and familiar with the technical nature of the disputes. The move in Canada in recent years has been to recognize the advantages of arbitrating construction disputes, a move that is overdue in comparison with arbitration approaches long established elsewhere through organizations such as the International Chamber of Commerce ("ICC") Court of Arbitration in Paris (established in 1923).

Popular support of international arbitration is evidenced by the adoption by the General Assembly of the United Nations of the UNCITRAL Arbitration Rules, rules established by the United Nations Commission on International Trade Law in 1976. Further evidence of popular support for international arbitration is evidenced by ratification in more than 80 countries of the New York Convention, enabling arbitral awards to be recognized and enforced in the more than 80 ratifying countries, including Canada.

The focus of this chapter is not on the details of the arbitration process, other than to support and recommend the implementation of the arbitration process as an advisable final and binding alternative to construction litigation. The ADR trends that are a central focus of this chapter relate primarily to pre-arbitral ADR measures that are designed to promote cooperative and contractual resolution of disputes. Such measures may in fact preclude the need for either the arbitration or litigation processes.

APPROACHES TO DISPUTE RESOLUTION ON SOME INTERNATIONAL PROJECTS

1. The Northumberland Strait Crossing Project

An example of a non-traditional approach to dispute resolution on a major Canadian infrastructure project is provided by the development contract for the P.E.I. bridge mega project, the Northumberland Strait Crossing Project. In that contract, resolution of technical disputes can be by way of binding arbitration, before a single agreed arbitrator or an appropriate panel of a technically qualified dispute review board.

The application of some binding arbitration decisions of dispute resolution boards on the Northumberland Strait Crossing Project is indicative of a trend that may receive further application on future projects, particularly given the reasonable concern of business participants in the construction industry to finally resolve disputes at the earliest possible stage.

2. The Boston Central Artery Tunnel Project

The Boston Central Artery Tunnel Project is a very substantial mega project involving some of the most complex highway projects ever undertaken. The project cost is expected to be well in excess of U.S. $6 billion. The construction schedule will span more than ten years. Located in the centre of downtown Boston in the centre of the City's business district, the project includes a four-lane tunnel under Boston Harbour, three complex major highway interchanges within Boston City limits, a new highway to be constructed under ten active railroad tracks, a depressed multi-lane artery throughout the centre of Boston, and a billion dollar bridge located in the centre of a multiple highway interchange on the northern side of the City.

It is expected that approximately one hundred construction contracts will be let on the project, some of the larger of which will be in the U.S. $200-300 million range. Also involved are several hundred subcontracts as well as contracts for construction management, preliminary and final design, and geo-technical services. Obviously, the project is immense and complicated.

Participation by Dispute Review Boards is contemplated for larger contracts on the Boston Central Artery Project. As pointed out in a 1993 presentation by James J. Myers, of the Gadsby & Hannah law firm in Boston:

> The Dispute Review Boards are formed by each of the contractors nominating a member (who can be from the contractor's organization, but not directly involved in the management or administration of this contract) and the Department nominating a member. The third member is to be impartial and is selected from lists of three nominees from each of the Department and the contractor.
>
> Within thirty (30) days after the authorized representative's determination of a dispute, any aggrieved party may file written demand for a hearing before the DRB. The other party has fourteen days to file an answer to the claim. Thereafter the DRB hears the dispute and has sole discretion to determine the manner of the hearing. Within sixty (60) days after the hearings are closed, the chairperson must submit written findings and recommendation to the Mass. Highway Commissioner. The Commissioner will review the DRB decision and will issue a written decision. For disputes under

$100,000 the Commissioner's decision will be "final and binding." For disputes over $100,000 the Commissioner's decision shall only be final if the contractor fails to protest the decision within twenty-one (21) days or within such time proceeds to litigation. Submittal of the claim to the DRB and the decision by the Mass. Highway Commission is a condition precedent to the filing of a protest with the Department's Hearing Examiner or for litigation in a court of law.

It is interesting that partnering, notwithstanding that it is voluntary, is addressed as part of the specifications on the Boston Central Artery Project. The statement from the specifications introducing the partnering provisions is as follows:

> The Department and the Management Consultant intend to encourage the foundation of a cohesive partnership with the Contractor and its Subcontractors. The partnership will be structured to draw on the strengths of each organization to identify and achieve reciprocal goals. The objectives include: effective and efficient contract performance; and completion with project within budget, on schedule, and in accordance with the plans and specifications.

3. Hong Kong Airport

The construction of Hong Kong's new airport is divided into two parts: the airport itself, which is the responsibility of the Provisional Authority, and the approaches to the Airport, which are the responsibility of the Hong Kong Government Works Branch. The airport and the approaches are collectively known as the Airport Core Programme. The project is immense. Construction cost estimates for the project range between 14 and 20 billion dollars U.S.

As pointed out by Anthony Connerty, a London barrister, in a paper he prepared for presentation in China in April, 1994, ADR techniques are evident in the government contracts for the Airport Core Programme in Hong Kong. The contracts contemplate four levels including: an initial decision on a dispute by the engineer; mediation; adjudication (a term describing a procedure much more formal than mediation but nevertheless a preliminary adjudication); and arbitration, as the final step in the process.

CONCLUSION

Implementation in Canada of the ADR techniques discussed, including partnering, mediation, dispute review boards, and arbitration makes a great deal of sense given the history of costly, time-consuming, and disruptive construction litigation in North America.

CHAPTER THIRTY

LIEN LEGISLATION

Every engineer in construction should be aware of provincial legislation that creates certain lien rights and that requires amounts to be held back from contractors until a specified time. In Ontario, the act is called The Construction Lien Act of Ontario. The Act significantly amended the old Mechanics' Lien Act of Ontario. The other common-law provinces also have lien legislation. In Alberta, British Columbia, and Manitoba, for example, the statute is called the Builders' Lien Act. In other common-law provinces the statute is called the Mechanics' Lien Act. In the Yukon and Northwest Territories it is called the Mechanics' Lien Ordinance. Engineers are often administrators of construction contracts; it is important that the engineer be aware of the necessity to comply with lien legislation.

DIFFERENCES IN PROVINCIAL LIEN ACTS

The general purpose of the various mechanics', construction, and builders' lien acts of the provinces is the same. But differences arise from province to province. Some of these differences concern matters of key significance to the engineer/contract-administrator. For example, calculation of the percentage holdback and the time of releasing holdback funds varies from province to province. The engineer-administrator should be familiar with applicable statutory requirements in the province in which the construction is taking place. The following table indicates, very generally, the respective percentage holdbacks and time periods in the various provinces after completion or abandonment of the contract, or otherwise as provided by the lien statute, before the holdback can be released. (It is extremely important to refer to the specific provincial statutes for more complete details, particularly in connection with the time period during which lien claims can be made and the timing of holdback release.)

	Percentage Holdback	Time before Holdback Can Be Released
Alberta	15%	45 days
British Columbia	10%	40 days
Manitoba	7.5%	40 days
Ontario	10%	45 days (Note: Ontario has two holdbacks, "basic" and "finishing")
New Brunswick	20%, if contract price doesn't exceed $15,000	60 days
	15%, if contract price exceeds $15,000 (a minimum of $3,000 must be held back)	60 days
Newfoundland	10%	30 days
Northwest Territories	10%	45 days
Nova Scotia	10%	45 days
Prince Edward Island	20%, if contract price doesn't exceed $15,000	60 days
	15%, if contract price exceeds $15,000 (a minimum of $3,000 must be held back)	60 days
Saskatchewan	10%	40 days
Yukon	10%	30 days

Essentially, until any lien is filed or until the owner is given notice in writing that a lien is claimed, an owner is protected if he or she retains the required percentage holdback of the value of the work done or materials furnished as construction proceeds. If an owner releases the holdback too soon, the owner may find it necessary to pay additional funds to satisfy lien claimants. Hence, knowledge of and compliance with the applicable lien act is very important.

PERSONS ENTITLED TO LIEN RIGHTS

The Construction Lien Act of Ontario provides, in effect, that anyone who supplies services or materials to an "improvement" for an owner, contractor, or subcontractor is entitled to a lien. The term "improvement," as defined in the Construction Lien Act, means:

i. any alteration, addition or repair to, or
ii. any construction, erection or installation on, any land, and includes the demolition or removal of any building, structure or works or part thereof, and "improved" has a corresponding meaning.

In Alberta, the Builders' Lien Act provides the following definition:

"improvement" means anything constructed, erected, built, placed, dug or drilled, or intended to be constructed, erected, built, placed, dug or drilled, on or in land except a thing that is neither affixed to the land nor intended to be or become part of the land.

RIGHTS AGAINST OWNER WHERE NO CONTRACT EXISTS

Construction lien rights exist in addition to any other rights that may exist at law. Privity of contract normally exists between an owner and a general contractor, but not between an owner and a subcontractor. Thus the lien rights provide the subcontractor with a cause of action directly against the owner. The subcontractor (or supplier or worker) would otherwise be limited in an action for breach of contract against the contractor. The lien statutes were designed to provide some degree of protection for parties who do not have privity of contract with the owner by providing a right to sue the owner directly under the statute. They also provide a specialized forum for resolution of payment disputes that arise during the course of construction projects.

EFFECT OF LIEN

The practical impact of a lien claim on a construction project can be most significant. In Ontario, when an owner receives written notice of a lien claim, the owner is obligated to retain both the holdback amount and the amount of the lien claim. Lien legislation in the other common-law provinces has the same effect: to generally ensure the retention of both the holdback amount and the amount of the lien claim. If an owner does not comply with the statutory requirements to hold back, and if a lien is filed, the owner may be liable to pay additional funds to satisfy the lien claim. Ultimately, the court might order the owner to sell its interest in the land to satisfy the lien claim if the owner is unable to or refuses to otherwise pay the amount of the claim. (Crown lands, however, may not be sold to satisfy lien claims.)

Lien legislation is admittedly complex. The engineer, as contract administrator, should obtain legal advice upon receiving notice of any lien claim. The owner might receive no written notice of lien claims during the course of the construction project. The engineer should still obtain legal advice to ensure that he or she can document compliance with the lien statute, to confirm the date at which holdback monies may be released, and to arrange to conduct a search at the appropriate registry office or land-titles office, to ensure that no lien claims have been registered before the engineer releases any portion of the holdback fund.

An engineer may be required to file a lien claim. For example, the engineer may be authorized to do so by a contractor in connection with a particular project. The engineer should obtain legal advice to ensure that the claim is made within the statutory-limitation period, and that it is otherwise in compliance with the applicable provincial lien statute.

THE ONTARIO CONSTRUCTION LIEN ACT

Some of the major aspects of The Construction Lien Act of Ontario (the "Act") that may be of particular interest to engineers are:

(a) **The holdback amount** — Pursuant to the Act, the amount of the holdback is ten percent of the price of services or materials as they are supplied. The Act creates two holdbacks. The first, or basic, holdback, is for work or services performed before it is certified that the contract is substantially performed. The second, or holdback for finishing work, is designed to give the finishing trades a claim against ten percent of the value of the remainder of the contract for the services and materials supplied from the date of substantial performance to the date the contract is completed. Because there are two separate holdbacks, lien claimants who want to make a claim for work done before the contract is substantially performed and for work done after it is done must file separate liens for such work.

(b) **Release of holdback** — Under the Act, holdback may be released when all liens that may be claimed against that holdback have expired or have been satisfied, discharged, or provided for by payment into court, under the provisions of the Act.

The rules for determining the period of time for retaining holdback are intended to provide certainty with respect to the date for holdback release. Note that the Act states times at which the payer *may* make payment of the remaining ninety percent of the contract and the holdback without jeopardy, and does not state that the payer *must* pay at such times. Therefore, there is no obligation to release holdback at a certain time under the Act; however, the obligation to release holdback funds is usually covered in the construction contract or subcontract.

The Act provides for certification of the *completion* of a subcontract to allow for the release of holdbacks related to a certified complete subcontract. Note that, pursuant to the Act, the determination of whether a subcontract has been completed may be made, upon the contractor's request, by the payment certifier (for example, the engineer) or by the owner and contractor jointly.

(c) **Certification of Substantial Performance** — There are statutory provisions dealing with certification or declaration of substantial performance of the contract, and for certification of completion (rather than substantial performance) of a subcontract. These concepts are important. They are two of the trigger points that determine when holdbacks can safely be paid out and when lien rights will expire. The definition of "substantial performance" in the Act is as follows:

2. — (1) For the purposes of this Act, a contract is substantially performed,
(a) when the improvement to be made under that contract or a substantial part thereof is ready for use or is being used for the purposes intended; and

(b) when the improvement to be made under that contract is capable of completion or, where there is a known defect, correction, at a cost of not more than,

(i)   3 per cent of the first $500,000 of the contract price,

(ii)   2 per cent of the next $500,000 of the contract price, and

(iii)   1 per cent of the balance of the contract price.

The Act also defines when a contract is deemed to be completed. Section 2(3) of the Act provides:

2. — (3) For the purposes of this Act, a contract shall be deemed to be completed and services or materials shall be deemed to be last supplied to the improvement when the price of completion, correction of a known defect or last supply is not more than the lesser of,

(a) 1 per cent of the contract price; and

(b) $1,000.

The rules relating to certification or declaration that a contract is substantially performed are mandatory; once they are invoked by the contractor, the parties must comply with the rules. On the application for certification made by the contractor, the payment certifier (usually an architect or engineer) or, if there is no payment certifier on the project, the owner and contractor together, must determine whether the contract has been substantially performed, and a certificate in the prescribed form must be signed. Once the contract is certified, the payment certifier, if any, must, within seven days of the date of executing the certificate, give or send a copy to the owner and the contractor. The contractor must publish a copy of the certificate in a "construction trade newspaper" as defined in the Act (for example, the *Daily Commercial News*). If the contractor does not do so, anyone may publish a copy of the certificate.

If the contract is not certified as substantially performed, any person may apply to the court and the court may declare the contract to be substantially performed. The person who brings the application must publish a copy of the court declaration in a "construction trade newspaper."

(d) **Damages for Non-Certification** — If an engineer doesn't certify substantial performance, the engineer may be liable for damages. The Act provides that anyone who is required to certify substantial performance (often, an engineer designated by the contract) and fails to do so within a reasonable time — even though there is no reasonable doubt that the contract has been substantially performed — will be liable to anyone who suffers damages as a result. In addition, a certifier who fails to deliver the certificate to the owner and contractor within seven days of certification will also be liable to anyone who suffers damage as a result.

Everyone should closely monitor the progress of a project, and check the construction trade newspapers regularly. Under the Act, the contractor must disclose, upon request in writing and within a reasonable time, the date of publication of certificates.

(e) **Who May Lien** — As noted earlier in this chapter, any person who supplies services or materials to an improvement will have a lien upon the property interest of the owner in the premises. The Act makes it clear that an "improvement" for which a person can claim a lien may include demolition or removal. In addition, the definition of "premises" and the definition of "materials" indicate that the lien applies not only to the improvement and the land upon which the improvement is made, but also to materials supplied to the improvement but not yet incorporated into the improvement.

The Act also provides that, where the construction of the improvement has not commenced, anyone (for example, an engineer) who supplies designs, plans, drawings, or specifications that enhance the value of the land is entitled to a lien. However, architects do not have lien rights. Liens by architects have been specifically excluded from the Act (architects pursued that exclusion). Careful consideration may have to be given to the appropriateness of retaining holdbacks from amounts paid to architects, for example, where engineers or others are retained as subconsultants. Even though architects can't file lien claims, and therefore will argue that the owner shouldn't retain a holdback on account of architectural services, architects' subconsultants may well have lien rights against the owner.

(f) **Prohibition Against Waiver of Lien Rights** — The Act provides that an agreement to waive lien rights is void.

(g) **Limits on Amount of Lien Claim** — Under the Act, a lien may be claimed only for the price of services and materials supplied prior to the time of the claim. The purpose of this provision is to reduce exaggerated lien claims. In addition, the Act provides that anyone making a frivolous or fictitious claim for lien will be liable to any person who suffers damage as a result.

(h) **Preservation of Lien Claims** — The term "preservation" is used in the Act. It is a term to describe the method of claiming a lien. The Act allows forty-five days. When the forty-five-day period commences depends on the circumstances. Bear in mind that there are two holdbacks (the basic holdback and the finishing holdback) and that different liens may apply to each. (If an engineer is planning to preserve a lien, it is advisable to have a lawyer do so.)

The following is a general summation of the time limits, under the Act, that apply to determine when the lien must be preserved:

(I) The lien of a contractor (as opposed to a subcontractor) for services or materials supplied before the date certified or declared to be the date of substantial performance must be preserved before the end of the forty-five-day period following the date of publication of substantial performance or the date the contract is either completed or abandoned — whichever is earlier.

(II) Where there is no certification or declaration of substantial performance, or for services or materials supplied after the certified date of substantial performance, the lien of a contractor must be preserved before the end of the forty-five-day period following the date the contract is completed or abandoned, whichever occurs first.

(III) The lien of any other person (subcontractors, suppliers and so on) for services or materials supplied on and before the certified date of substantial performance of the contract must be preserved before the end of the forty-five-day period next following the earliest of:

- (i) the date of publishing the certificate or declaration of substantial performance of the contract;
- (ii) the date on which he last supplies services or materials to the improvement; and
- (iii) the date a subcontract is certified to be complete (where the services or materials were supplied under that subcontract).

(IV) Where there is no certification or declaration of substantial performance of the contract, or for services or materials supplied after the certified date of substantial performance of the contract, the lien of any person other than the contractor must be preserved before the end of the forty-five-day period following the date on which he last supplies services or materials to the improvement and the date a subcontract is certified to be complete (where the services or materials were supplied under that subcontract) — whichever is earlier.

As with any lawsuit, the engineer should understand that a lawyer ought to be retained to advise and act on behalf of the engineer or the engineer's client in any lien action.

(i) **Priority of Mortgages** — Many readers may be aware of certain difficulties that lending institutions — such as banks and trust companies — have raised with the Act, particularly because a lien claimant is given priority over the lender in certain circumstances (to the extent of any deficiency in the holdbacks required by the Act).

The Act distinguishes between three classes of mortgages: the building mortgage (that is, a mortgage taken to finance the construction project); the prior mortgage (that is, any mortgage registered prior to the time a lien first arises); and the subsequent mortgage (that is, any mortgage registered after the time a lien first arises). The Act gives lien claimants priority over building mortgages taken out at the outset of the project, long before any lien might arise, insofar as holdback deficiencies are concerned.

However, the Act permits a financial-guarantee bond to be registered on title. If such a bond is registered, the liability for deficiencies in holdbacks will expire if the lender (the mortgagee) sells the property. This provision is designed to allow a mortgagee to sell the premises in the event of a default under the mortgage free of the liability for such deficiencies in holdbacks. Lien claimants can then claim against the surety (the bonding company) on the bond for such deficiencies.

THE TRUST FUND

Trust obligations are included in the Act. Suppose a contract states that the owner must pay the contractor a certain sum when an authorized person — an engineer — certifies. That sum constitutes a trust fund in the hands of the owner for the benefit of the contractor, and no part of the trust fund can be appropriated or converted for use by the owner except in accordance with the trust-fund provisions of the Construction Lien Act. Similar trust-fund provisions also extend to amounts received by contractors and subcontractors; such amounts are subject to a trust for the benefit of those who have supplied services or materials. The funds cannot be used otherwise until the specified beneficiaries have been paid in full.

Similar trust-fund provisions are included in mechanics' lien legislation in British Columbia, Saskatchewan, Manitoba, and New Brunswick.

In Ontario, new trust obligations under the Act include and extend to all amounts received by an owner after substantial performance, and to amounts received by an owner as a result of a sale of the premises (less reasonable expenses of the sale and amounts to discharge mortgages).

In the event of non-payment of trust monies, and where lien rights have expired, an action under the trust-fund provisions to recover misappropriated trust funds may provide a remedy.

ENGINEERS' RIGHTS TO LIEN CLAIMS

As already noted, pursuant to the Act in Ontario, engineers' entitlement to lien claims is now confirmed by statute, provided the engineers' services enhance the value of the land.

In the other common-law provinces, where the engineer's right to lien is not yet statutory, what constitutes "necessary services of engineers and architects" to qualify for lien claims remains a question to be decided by the courts of those provinces. There have already been a number of such court decisions.

In 1971, the British Columbia Supreme Court decided the *Application of Erickson/Massey*.[1] An architect prepared plans for, and supervised the construction of, a building. The court held that he was entitled to a lien under the Mechanics' Lien Act of British Columbia. An earlier decision held that an architect was not entitled to a claim for lien. But the earlier decision was distinguished: the architect had only prepared plans. He had not supervised the construction of the building.

But providing supervisory services is not necessarily a prerequisite for an engineer's entitlement to a lien in all provinces. For example, in 1965, the Supreme Court of Alberta decided *Englewood Plumbing & Gas Fitting Ltd.* v. *Northgate Development Ltd. et al.*[2] The court held that an architect was entitled to

[1]   [1971] 2 W.W.R. 767
[2]   [1966] 54 W.W.R. 225

file a lien claiming payment for his services in the preparation of plans, although the architect had performed no supervisory services in connection with the construction.

In 1975, the Supreme Court of Ontario decided *Armbro Materials and Construction Ltd.* v. *230056 Investments Limited et al.*[3] An engineer had prepared plans for sewers, water mains, and roads in a subdivision; the plans had to be approved by municipal authorities. The engineer was also retained to supervise construction of the services. Approval of the municipality was obtained, but construction did not proceed for financial reasons. The engineer made a claim for lien. The court distinguished the particular plans of the engineer from architectural plans, and noted that the engineering plans had improved the value of the land. The engineer was entitled to his claim for lien. The court stated:

> ... those engineering services are so inextricably linked with the land. They are different from the plans of an architect. The plans of an architect are, may I say, "up in the air," and they can fit everywhere, and if they do not fit that specific parcel, then they become a stock plan and they can be sold to someone else who has another lot, and therefore those plans do not need as much protection because they have a merchantable value. They can be sold to someone else if the original man who ordered the plans cannot pay for them, or does not need them any more, but those engineering plans are very specific. They are not "in the air" — they run with the land. The owner could sell those 17 acres now at a much higher price than he could before those plans were made and approved. It is the approval that gives a good deal of value to the lands because it is a long process and a difficult one but once it is obtained, it is attached to the lands and the purchaser can benefit with those plans immediately. He does not have to begin all over again. He carries on from there.

> Therefore, I find that those plans, approved as they are, advance the value of the land considerably and they are specifically for that parcel of land and therefore I think it would be unjust and unfair not to grant a lien to the people who have increased the value of the land.

[3] 9 O.R. (2d) 226

LIEN STATUTES:

Alberta: The Builders' Lien Act, R.S.A. 1980, c. B-12

British Columbia: The Builders' Lien Act, R.S.B.C. 1979, c. 40

Manitoba: The Builders' Liens Act, R.S.M. 1987 c. B 91

New Brunswick: The Mechanics' Lien Act, R.S.N.B. 1973, c. M-6

Newfoundland: The Mechanics' Lien Act, R.S.N. 1990 c. M-3

Nova Scotia: The Mechanics' Lien Act, R.S.N.S. 1989 c. 277

Ontario: The Construction Lien Act, R.S.O. 1990, c. C. 30

Prince Edward Island: The Mechanics' Lien Act, R.S.P.E.I. 1988, c. M-5

Saskatchewan: The Mechanics' Lien Act, R.S.S. 1978, c. M-7

Yukon: Mechanics' Lien Act RSY 1986, c. 112

North West Territories: Mechanics' Lien Act R.S.N.W.T. 1988, c. M-7

CHAPTER THIRTY-ONE

THE COMPETITION ACT

The *Competition Act*[1] is a federal statute designed to maintain and encourage business competition in Canada; to promote the efficiency of the Canadian economy; and to expand opportunities for Canadian participation in world markets. The Act prescribes certain offences and very significant penalties. The scope of offences under the Act is broad: it deals, for example, with formation of monopolies, misleading advertising, bid-rigging, price fixing, and conspiracy to unduly limit competition.

An increasing number of convictions, particularly in the area of misleading advertising, attests to the Act's significance. "White collar crime" is undoubtedly a "high-profile" matter today; the engineer in business should be aware of the Act. Its broad scope is perhaps best illustrated through examining selected sections of the statute.

MISLEADING ADVERTISING

Section 52(1), which deals with misleading advertising, provides:

> No person shall, for the purpose of promoting, directly or indirectly, the supply or use of a product or for the purpose of promoting, directly or indirectly, any business interest, by any means whatever,
> (a) make a representation to the public that is false or misleading in a material respect;
> (b) make a representation to the public in the form of a statement, warranty or guarantee of the performance, efficacy or length of life of a product that is not based on adequate and proper test thereof, the proof of which lies upon the person making the representation;

[1] R.S.C. 1985, c. C-34

(c) make a representation to the public in a form that purports to be
   (i) a warranty or guarantee of a product, or
   (ii) a promise to replace, maintain or repair an article or any part thereof or to repeat or continue a service until it has achieved a specified result

if the form of purported warranty or guarantee or promise is materially misleading or if there is no reasonable prospect that it will be carried out; or

(d) make a materially misleading representation to the public concerning the price at which a product or like products have been, are or will be ordinarily sold, and for the purposes of this paragraph a representation as to price is deemed to refer to the price at which the product has been sold by sellers generally in the relevant market unless it is clearly specified to be the price at which the product has been sold by the person by whom or on whose behalf the representation is made.

Conviction can result in a fine "in the discretion of the Court" — that is, there is no maximum limit — or to imprisonment for five years, or to both.

BID-RIGGING

The offence known as "bid-rigging" (that is, collusion leading to the submission of fraudulent contract tenders or bids) is defined in section 47(1) of the Act as follows:

In this section "bid-rigging" means

(a) an agreement or arrangement between or among two or more persons whereby one or more of such persons agrees or undertakes not to submit a bid in response to a call or request for bids or tenders, or

(b) the submission, in response to a call or request for bids or tenders, of bids or tenders that are arrived at by agreement or arrangement between or among two or more bidders or tenderers,

where the agreement or arrangement is not made known to the person calling for or requesting the bids or tenders at or before the time when any bid or tender is made by any person who is a party to the agreement or arrangement. Section 47(2) provides:

Every one who is a party to bid-rigging is guilty of an indictable offence and liable on conviction to a fine in the discretion of the court or to imprisonment for a term not exceeding five years or to both.

CONSPIRACY

The section of the Act that deals with conspiracy (and focuses on preventing interference with competition) is very broadly worded. Section 45(1) provides:

Every one who conspires, combines, agrees or arranges with another person

(a) to limit unduly the facilities for transporting, producing, manufacturing, supplying, storing or dealing in any product,

(b) to prevent, limit or lessen, unduly, the manufacture or production of a product, or to enhance unreasonably the price thereof,

(c) to prevent, or lessen, unduly, competition in the production, manufacture, purchase, barter, sale, storage, rental, transportation or supply of a product, or in the price of insurance on persons or property, or

(d) to otherwise restrain or injure competition unduly,

is guilty of an indictable offence and liable to imprisonment for a term not exceeding five years or to a fine not exceeding ten million dollars or to both.

The court must determine in each case, whether the "lessening of competition," for example, is undue; interpreting the meaning of "undue" in any particular circumstances may provide considerable latitude for argument.

TRADE ASSOCIATIONS

Certain activities of trade associations — associations of manufacturers and contractors within particular industries — are permitted under the Act. Pursuant to section 45(3) of the Act, the permitted activities or agreements must relate to the following subjects:

(a) the exchange of statistics

(b) the defining of product standards

(c) the exchange of credit information

(d) the definition of terminology used in a trade, industry, or profession

(e) cooperation in research and development

(f) the restriction of advertising or promotion, other than a discriminatory restriction directed against a member of the mass media

(g) the sizes or shapes of the containers in which an article is packaged

(h) the adoption of the metric system of weights and measures

(i) measures to protect the environment

The activities listed are legitimate subjects of exchange between trade-association members only insofar as no agreement or arrangement is entered into that will lessen competition unduly in respect of prices, quantity, or quality of production, markets or customers, or channels or methods of distribution; or which would restrict any person from entering into or expanding a business in the trade or industry. An exchange of statistics, for example, might appear to be an inherently innocent activity, but it could cross the fine line of illegality. Suppose, for example, that a statistical table of costs were used by members of a trade association as a basis for standard mark-ups; that use could lead to price-fixing. (*Competitors should strictly avoid any discussion of prices.*) As well, technical discussions within an association that discourage independent research by competing companies, or that establish industry standards that smaller, newer, or potential competitors may not be able to meet, may constitute an offence under the Act.

Association members must avoid discussion or other activity that would allocate market shares among firms, allocate markets in any geographical sense, or allocate customers in any manner. An understanding that would limit transportation facilities or restrict channels or methods of distribution should be avoided.

Associations must be most cautious not to reach any understanding or pursue any activity (such as an agreement to refuse to deal) that would lessen competition unduly by restricting any person from entering into a particular business or industry.

CHAPTER THIRTY-TWO

REGULATORY ASPECTS AND ETHICS

Each of Canada's provinces and territories has enacted legislation to govern the practice of professional engineering. The professional engineer should become acquainted with applicable legislation in the province in which he or she practises.

PURPOSE OF LEGISLATION

The general purpose of the governing legislation is to regulate the practice of professional engineering, to protect the public interest. For example, sections 2(3) and 2(4) of the Professional Engineers Act[1] of Ontario provide:

> (3) The principal object of the Association is to regulate the practice of professional engineering and to govern its members, holders of certificates of authorization, holders of temporary licences and holders of limited licences in accordance with this Act, the regulations and the by-laws in order that the public interest may be served and protected.
> (4) For the purpose of carrying out its principal object, the Association has the following additional objects:
>> 1. To establish, maintain and develop standards of knowledge and skill among its members.
>> 2. To establish, maintain and develop standards of qualification and standards of practice for the practice of professional engineering.
>> 3. To establish, maintain and develop standards of professional ethics among its members.

[1] R.S.O., 1990, c.P.28

4. To promote public awareness of the role of the Association.

5. To perform such other duties and exercise such other powers as are imposed or conferred on the Association by or under any Act.

The regulations under the Professional Engineers Act of Ontario respond to the objects of the Association by addressing a wide range of matters.[2] For example, the regulations detail the structure of PEO's Council and provide for such matters as procedures for election to Council; mediating disputed engineering fees; issuing licences and certificates of authorization; prescribing requisite examinations; designating specialists and consulting engineers; defining "professional misconduct"; and setting out minimum requirements for professional liability insurance; as well as specifying the Code of Ethics of Professional Engineers Ontario ("PEO").

DEFINITION OF PROFESSIONAL ENGINEERING

"Professional engineering" is defined in the various provincial statutes. The definitions are broad in scope. The "practice of professional engineering" is defined in section 1 of the Professional Engineers Act of Ontario:

> "practice of professional engineering" means any act of designing, composing, evaluating, advising, reporting, directing or supervising wherein the safeguarding of life, health, property or the public welfare is concerned and that requires the application of engineering principles, but does not include practising as a natural scientist.

"Practice of engineering" is defined in Section 1(m) of the Engineering, Geological and Geophysical Professions Act of Alberta[3]:

> "practice of engineering" means
> (i) reporting on, advising on, evaluating, designing, preparing plans and specifications for or directing the construction, technical inspection, maintenance or operation of any structure, work or process

[2] O. Reg. 538/84
[3] S.A. Chap. E-11.1

(A) that is aimed at the discovery, development or utilization of matter, materials or energy or in any other way designed for the use and convenience of man, and

(B) that requires in the reporting, advising, evaluating, designing, preparation or direction the professional application of the principles of mathematics, chemistry, physics or any related applied subject, or

(ii) teaching engineering at a university.

PROFESSIONAL ENGINEER'S SEAL

In all of the common-law jurisdictions in Canada, a professional engineer is required to stamp drawings and specifications with his or her seal. The seal is issued by the Provincial Association, and it indicates that the engineer is a registered Professional Engineer. It is extremely important that each engineer closely controls the use of that seal and ensures that it is only used where appropriate. Improper use of the engineer's seal in Ontario can result in disciplinary proceedings and very substantial fines for individuals, corporations, and partnerships. In Ontario, the Council of Professional Engineers Ontario is empowered to make regulations "requiring and governing the signing and sealing of documents and designs by members of the Association, holders of temporary licences and holders of limited licences, specifying the forms of seals and respecting the issuance and ownership of seals." And Section 53 of the regulations under the Professional Engineers Act provides:

> Every holder of a licence, temporary licence or limited licence who provides to the public a service that is within the practice of professional engineering shall sign, date and affix the holder's seal to every final drawing, specification, plan, report or other document prepared or checked by the holder as part of the service before it is issued.

The appropriate use of the engineer's seal is extremely important. Section 72(2)(e) of the Ontario regulations includes, as part of the definition of "professional misconduct":

(e) signing or sealing a final drawing, specification, plan, report or other document not actually prepared or checked by the practitioner.

In Alberta, a professional engineer is required to use his or her seal pursuant to section 76 of the Engineering, Geological and Geophysical Professions Act of Alberta and Part 11 of the Regulations thereunder. Section 76 provides that:

> 76(1) A professional member, license or restricted practitioner shall sign and stamp or seal all documents or records in accordance with the regulations.
>
> (2) No person other than a professional member, licensee, permit holder or certificate holder shall use a stamp or seal issued by the Registrar under this Act.

PARTNERSHIPS AND CORPORATIONS

The regulatory provincial common-law statutes provide, in effect, that a partnership or corporation may be granted a certificate of authorization or be permitted to practise professional engineering, providing such professional engineering is practised under the responsibility and supervision of a member or of a licensed professional engineer and otherwise complies with the applicable provincial statute and regulations thereunder.

DISCIPLINARY HEARINGS

The regulatory statutes also authorize what disciplinary action may be taken against members or licensees for professional misconduct. Disciplinary action usually arises, in Ontario, for example, following the submission of a complaint to the Complaints Committee. The Committee examines the complaint and can then refer the matter to the Discipline Committee. The Discipline Committee is authorized (pursuant to section 28 of the Professional Engineers Act), when so directed by the Council, the Executive Committee or the Complaints Committee, to hear and determine allegations of professional misconduct or incompetence. Disciplinary action may result in reprimands, suspensions, fines, and cancellation of membership and licences. Decisions of disciplinary hearings may be appealed to the courts in accordance with the provisions of the applicable regulatory statute.

PENALTIES

The offence provisions of the statutes that regulate engineering impose varying penalties for contravening statutes and ordinances. The penalty in Ontario, for example, for practising professional engineering without a licence or for holding oneself out as engaging in the practice of engineering without being properly licensed includes a fine of not more than $25,000 for the first offence, and for each subsequent offence a fine of not more than $50,000.

CERTIFICATES OF AUTHORIZATION

In Ontario, professional engineering membership alone does not qualify engineers to offer to the public, or engage in the business of providing to the public, services that are within the practice of professional engineering. A certificate of authorization is also required. Applicants for certificates of authorization, including individual members, partnerships, and corporations, must meet prescribed requirements and qualifications pursuant to the Professional Engineers Act. The registrar is authorized to refuse to issue, to suspend, or to revoke a certificate of authorization where the registrar "is of the opinion, upon reasonable and probable grounds" that the certificate should be refused, suspended, or revoked, upon the basis of past conduct or failure to meet requirements or qualifications.

All holders of certificates of authorization must have professional liability insurance, subject, however to certain exceptions pursuant to Section 74 of the regulations. Minimum coverage for each single occurrence must be not less than $250,000, with an aggregate policy limit for all occurrences of not less than $500,000 per year or, alternatively, an automatic-policy-limit-reinstatement feature. Any engineer who obtains such insurance should ensure that it complies with all the requirements of the regulations. Section 74(2) of the regulations provides for certain circumstances in which a holder of a certificate of authorization is not required to be insured, including circumstances where insurance is not available (e.g., nuclear hazards) and where written authority from the client for professional engineering services to be performed without insurance is provided. Regulations, like statutes, may change from time to time. It is most important for professional engineers to keep informed of current professional liability insurance requirements under their governing provincial statutes and regulations.

OVERLAPPING IN THE SCOPE OF ENGINEERING AND ARCHITECTURAL PRACTICES

As noted, the term "professional engineering" is broadly defined by the regulatory statutes and ordinances. Occasionally, the broad definitions lead to confusion; one problem is the distinction between the functions of an engineer and an architect. The problem has been considered by the courts.

For example, in 1939 the court heard *Rex* v. *Bentall.*[4] A professional engineer in British Columbia had planned and supervised the construction of a theatre; he was convicted for unlawfully practising as an architect.

In 1955, the Ontario County Court heard *Regina* v. *Margison and Associates, Limited.*[5] An engineering firm was charged with holding itself out as an architect. The court examined both the Professional Engineers Act and the Architects Act to decide whether a particular job was essentially one for an engineer or for an architect. But neither Act provided the court with definitions specific enough to distinguish between the two professions. The action against the engineer failed. The court stated, in part:

> I cannot but think that it was the intention of the legislation to give reciprocal privileges, at least to the extent necessary to cover the facts disclosed, and that it is up to a client to weigh the qualifications of firms of architects and engineers and decide which he wishes to employ, or indeed if he wishes to employ both, which is a common practice. Mr. Fleming contends that it is possible to decide whether a job is essentially one for an architect or one for an engineer. I can find nothing in the words of the legislation which enables the Court to draw such a line between the two professions, and I do not think that the Court should endeavour to do so until the Legislature passes appropriate provisions.

[4]   [1939] 3 W.W.R. 39
[5]   [1955] O.W.N. 705

BACKGROUND TO CHANGES IN ONTARIO LEGISLATION

In Ontario, however, the Professional Engineers Act includes provisions intended to simplify the "scope-of-practice" dispute between professional engineers and architects. The provisions describe the nature of construction projects that can be designed by engineers alone, or by architects alone, or by both.

The provisions were worked out by representatives of the PEO and the Ontario Association of Architects (OAA). Premised on two particular declarations of principle, agreement was reached. The first premise: engineers should confine their professional activity to the practice of engineering and architects to the practice of architecture. The second premise: a client should be free to select the prime consultant of the client's choice. The PEO–OAA agreement and the decision of the Professional Organizations Committee contemplated in part that a Joint Practice Board, composed of three engineers and three architects appointed by PEO and the OAA respectively, be instituted. The Board would be authorized to make recommendations to the PEO and the OAA and to handle complaints of an interprofessional nature. It would also be empowered to work on other matters, such as the co-ordination and publication of guidelines, standards, criteria, and performance standards in the field of building design and construction.

The recommendations of the Professional Organizations Committee were implemented and resulted in revisions to regulatory statutes governing both engineers and architects. Section 47 of the Professional Engineers Act of Ontario now provides for a Joint Practice Board composed of a chair, three architects, and three professional engineers, who are authorized to recommend the issuance of licences. Firms are now entitled to practise both professional engineering and architecture.

The Joint Practice Board is also authorized to assist in the resolution of "scope-of-practice" disputes. Sections 47(4) and 47(5) of the Professional Engineers Act provide:

> 47(4) Where a dispute arises between an architect and a professional engineer or a holder of a certificate of authorization as to jurisdiction in respect of professional services, the Registrar may refer the matter to the Joint Practice Board and the Joint Practice Board shall consider the matter and assist the

architect and the professional engineer or the holder of the certificate of authorization to resolve the dispute in accordance with the rules in section 12.

(5) Proceedings shall not be commenced under this Act in respect of a matter mentioned in subsection (4) except upon the certificate of the chair of the Joint Practice Board that the Board has considered the matter and has been unable to resolve the dispute.

LITIGATION VS. DISCIPLINARY MATTERS

As previously explained in this text, in order for a professional engineer to be liable for tort remedies or for the remedy of damages for breach of a contract involving a duty of care (such as a contract for engineering services) it is necessary for damages to have been caused by the engineer's negligence, whether the lawsuit is in tort or contract or, as is more likely to be the case, in both. To more briefly state the concept, damages having been caused by the negligent engineer are a prerequisite to the defendant engineer being liable in such a lawsuit. Such is not the case, however, as far as disciplinary proceedings under the Professional Engineer's Act are concerned. Damages are not a prerequisite in disciplinary proceedings.

Disciplinary proceedings under the Professional Engineer's Act usually proceed in the form of allegations of professional misconduct or incompetence. In this regard, section 72 of the Regulations under the Professional Engineer's Act sets out a definition of negligence and an extensive description of conduct that amounts to professional misconduct. Not surprisingly, "negligence" is defined to mean an act or an omission in the carrying out of the work of a practitioner that constitutes a failure to maintain the standards that a reasonable and prudent practitioner would maintain in the circumstances. Obviously the definition of negligence is tied to the measure of the duty of care that a court would apply in a lawsuit in determining if a professional engineer had been negligent. Accordingly, any conduct constituting negligence by tort standards will also constitute professional misconduct.

Accordingly, the disciplinary process under the Professional Engineer's Act provides for circumstances in which a professional engineer who has been negligent can be sanctioned even where damages have not been a consequence of the professional engineer's negligent act or omission. For example, where it could be substantiated that a professional engineer had negligently prepared drawings and specifications, the professional engineer could face disciplinary hearings even though construction of the structure described by the drawings and specifications had never taken place.

Professional engineering is a self-governing profession. The disciplinary process provides sanctions that can be applied independently of any lawsuit and is therefore a very important basis upon which the provincial bodies may govern, in the interests of maintaining professional standards to ensure that the public interest is served and protected.

THE CODE OF ETHICS

The Codes of Ethics under the Professional Engineers Act of Ontario (and in place in the other provinces and territories of Canada) and as endorsed by the Canadian Council of Professional Engineers provide for appropriately high standards of duty, conduct, and integrity. Such high standards are obviously extremely important from a technical perspective given the responsibility assumed by, and integral to, the undertakings of professional engineers as guardians of the public safety. It is extremely important that professional engineers respect and implement their codes of ethics as professionals discharging their duties to the public, their employers, their clients, their colleagues, their profession, and themselves. Engineers have long fulfilled important leadership roles in society and enjoy a high degree of respect because of the achievements, abilities, and integrity engineers have demonstrated. Maintaining a strong sense of duty and otherwise fulfilling the expectations of the Code of Ethics is also important to maintaining the strength and esteem of the engineering profession in society.

Section 77 of the regulations pursuant to the Professional Engineers Act of Ontario sets out PEO's Code of Ethics. The code contains standards of conduct designed for the protection of the public. Members and licensees must subscribe to and follow those standards in the practice of professional engineering. The code includes important duties each engineer owes to the public, to employers, to clients, to colleagues, and to the profession, as well as to the engineer himself or herself.

Every engineer should therefore be familiar with his or her provincial code of ethics. The Ontario Code of Ethics is as follows:

77. The following is the Code of Ethics of the Association:
1. It is the duty of a practitioner to the public, to the practitioner's employer, to the practitioner's clients, to other members of the practitioner's profession, and to the practitioner to act at all times with,
   i. fairness and loyalty to the practitioner's associates, employers, clients, subordinates and employees,
   ii. fidelity to public needs,
   iii. devotion to high ideals of personal honour and professional integrity,
   iv. knowledge of developments in the area of professional engineering relevant to any services that are undertaken, and
   v. competence in the performance of any professional engineering services that are undertaken.
2. A practitioner shall,
   i. regard the practitioner's duty to public welfare as paramount,
   ii. endeavour at all times to enhance the public regard for the practitioner's profession by extending the public knowledge thereof and discouraging untrue, unfair or exaggerated statements with respect to professional engineering,
   iii. not express publicly, or while the practitioner is serving as a witness before a court, commission or other tribunal, opinions on professional engineering matters that are not founded on adequate knowledge and honest conviction,
   iv. endeavour to keep the practitioner's licence, temporary licence, limited licence or certificate of authorization, as the case may be, permanently displayed in the practitioner's place of business.
3. A practitioner shall act in professional engineering matters for each employer as a faithful agent or trustee and shall regard as confidential information obtained by the practitioner as to the business affairs, technical methods or processes of an employer and avoid or disclose a conflict of interest that might influence the practitioner's actions or judgment.

4. A practitioner must disclose immediately to the practitioner's client any interest, direct or indirect, that might be construed as prejudicial in any way to the professional judgment of the practitioner in rendering service to the client.

5. A practitioner who is an employee-engineer and is contracting in the practitioner's own name to perform professional engineering work for other than the practitioner's employer, must provide the practitioner's client with a written statement of the nature of the practitioner's status as an employee and the attendant limitations on the practitioner's services to the client, must satisfy the practitioner that the work will not conflict with the practitioner's duty to the practitioner's employer, and must inform the practitioner's employer of the work.

6. A practitioner must co-operate in working with other professionals engaged on a project.

7. A practitioner shall,
    i.   act towards other practitioners with courtesy and good faith,
    ii. not accept an engagement to review the work of another practitioner for the same employer except with the knowledge of the other practitioner or except where the connection of the other practitioner with the work has been terminated,
    iii. not maliciously injure the reputation or business of another practitioner,
    iv. not attempt to gain an advantage over other practitioners by paying or accepting a commission in securing professional engineering work, and
    v.   give proper credit for engineering work, uphold the principle of adequate compensation for engineering work, provide opportunity for professional development and advancement of the practitioner's associates and subordinates, and extend the effectiveness of the profession through the interchange of engineering information and experience.

8. A practitioner shall maintain the honour and integrity of the practitioner's profession and without fear or favour expose before the proper tribunals unprofessional, dishonest or unethical conduct by any other practitioner.

The Code of Ethics is clearly stated. Obviously, it focuses on the importance of desirable characteristics that engineers readily understand, such as fairness, loyalty, honour and integrity. It clearly emphasizes the paramountcy of the engineer's duty to public welfare. The Code of Ethics also emphasizes the importance of protecting confidential information of employers; of disclosing conflicts of interest to clients; as well as informing clients if the professional engineer is otherwise engaged in an employment relationship. Further, the Code of Ethics emphasizes the importance of cooperating and otherwise respecting and fairly dealing with other colleagues.

Where a professional engineer undertakes to review the work of another professional engineer, the other professional engineer is to be so notified.

The responsibility of the engineer to the engineering profession is to maintain the profession's honour and integrity. To do so may require giving testimony to expose dishonest or unethical conduct by a colleague.

The important standards of the Code of Ethics are also reflected in the various descriptions of "professional misconduct" pursuant to section 72 of the Professional Engineers Act of Ontario. Section 72 sets out important specific definitions of negligence and professional misconduct. Section 72 provides:

> 72. — (1) In this section, "negligence" means an act or an omission in the carrying out of the work of a practitioner that constitutes a failure to maintain the standards that a reasonable and prudent practitioner would maintain in the circumstances.
>
> (2) For the purposes of the Act and this Regulation, "professional misconduct" means,
>
> (a) negligence;
>
> (b) failure to make reasonable provision for the safeguarding of life, health or property of a person who may be affected by the work for which the practitioner is responsible;
>
> (c) failure to act to correct or report a situation that the practitioner believes may endanger the safety or the welfare of the public;
>
> (d) failure to make responsible provision for complying with applicable statutes, regulations, standards, codes, by-laws and rules in connection with work being undertaken by or under the responsibility of the practitioner;

(e) signing or sealing a final drawing, specification, plan, report or other document not actually prepared or checked by the practitioner;

(f) failure of a practitioner to present clearly to his employer the consequences to be expected from a deviation proposed in work, if the professional engineering judgment of the practitioner is overruled by non-technical authority in cases where the practitioner is responsible for the technical adequacy of professional engineering work;

(g) breach of the Act or regulations, other than an action that is solely a breach of the Code of Ethics;

(h) undertaking work the practitioner is not competent to perform by virtue of his training and experience;

(i) failure to make prompt, voluntary and complete disclosure of an interest, direct or indirect, that might in any way be, or be construed as, prejudicial to the professional judgment of the practitioner in rendering service to the public, to an employer or to a client, and in particular without limiting the generality of the foregoing, carrying out any of the following acts without making such a prior disclosure:

1. Accepting compensation in any form for a particular service from more than one party.

2. Submitting a tender or acting as a contractor in respect of work upon which the practitioner may be performing as a professional engineer.

3. Participating in the supply of material or equipment to be used by the employer or client of the practitioner.

4. Contracting in the practitioner's own right to perform professional engineering services for other than the practitioner's employer.

5. Expressing opinions or making statements concerning matters within the practice of professional engineering of public interest where the opinions or statements are inspired or paid for by other interests;

(j) conduct or an act relevant to the practice of professional engineering that, having regard to all the circumstances, would reasonably be regarded by the engineering profession as disgraceful, dishonourable or unprofessional;

(k) failure by a practitioner to abide by the terms, conditions or limitations of the practitioner's licence, limited licence, temporary licence or certificate;

    (l) failure to supply documents or information requested by an investigator acting under section 34 of the Act;

    (m) permitting, counselling or assisting a person who is not a practitioner to engage in the practice of professional engineering except as provided for in the Act or the regulations.

Professional misconduct, as so described, has far-reaching potential. That potential needs to be carefully considered in the circumstances of each engineer's business engagements. Compliance with technical standards is obviously essential, as is practising only within one's competence. Failure to take action to safeguard life, health, or property or the safety or welfare of the public occupies a prominent place in the list of what constitutes professional misconduct, as does negligence. Failure to warn an employer of the consequences to be expected from a proposed deviation in the work, where the engineer is overruled by a non-technical authority, is a situation that can arise where "shortcuts" may be proposed by other project participants or by employers. The engineer must stay alert to such developments and issue appropriate warnings accordingly.

A professional misconduct item that clearly overlaps with the duties prescribed by the Code of Ethics is conduct giving rise to a "conflict of interest" pursuant to section 72(2)(i) of the Regulations. The section is helpful in setting out particular examples, all of which demonstrate circumstances in which there is a potential for the engineer to inappropriately personally gain. Disclosure of conflicts of interest are most important. Examples contemplated include taking secret commissions; bidding as a contractor on a portion of the project on which the engineer is providing professional services; acting as a supplier to a project on which the engineer is engaged; and contracting outside the scope of the engineer's employment without disclosure to the client.

The listing of professional misconduct items should be very helpful to the professional engineer in practice, provided that the engineer appreciates that the potential scope of the application of the principles reflected in the list is very broad. The professional engineer needs to be astute and alert to the high standards of the Code of Ethics and the definitions of professional misconduct set out in the Regulations.

CHAPTER THIRTY-THREE

INDUSTRIAL PROPERTY

Rights that relate generally to patents, trade-marks, copyrights, and industrial designs are sometimes called "industrial-property rights." The federal Patent Act,[1] the Trade-Marks Act,[2] the Copyright Act,[3] and the Industrial Design Act[4] govern these rights. The statutes describe such matters as obtaining patents of invention and the registration of trade-marks, copyrights, and industrial designs. Patent applications and trade-mark, industrial design, and copyright registrations are usually effected by professionals who specialize in such matters. This chapter will focus on the basic nature and practical aspects of industrial-property rights.

PATENTS OF INVENTION

Definition

The Patent Act defines an invention as "any new and useful art, process, machine, manufacture or composition of matter, or any new and useful improvement in any art, process, machine, manufacture or composition of matter."

What May Be Patented

What constitutes patentable subject matter? The definition of "invention" in the Patent Act is a broad one. The invention must be novel and useful. An idea alone is not patentable: the idea or principle must be reduced to something physical. The need for something tangible was discussed in *Permutit Co.* v. *Borrowman*[5]:

[1] R.S.C. 1985, c. P-4
[2] R.S.C. 1985, c. T-13
[3] R.S.C. 1985, c. C-42
[4] R.S.C. 1985, c. I-9
[5] [1926] 43 R.P.C. 356

It is not enough for a man to say that an idea floated through his brain; he must at least have reduced it to a definite and practical shape before he can be said to have invented a process.

In 1929, the Privy Council decided *General Electric Company, Limited* v. *Fada Radio, Limited.*[6] The following excerpt focusses on the idea that patentable inventions must have two characteristics — utility and novelty — that result from the application of ingenuity and skill.

> The law on this subject is, in their Lordships' opinion, accurately summarized by Maclean J. in his judgment. His statement is as follows: "There must be a substantial exercise of the inventive power or inventive genius, though it may in cases be very slight. Slight alterations or improvements may produce important results, and may disclose great ingenuity. Sometimes it is a combination that is the invention; if the invention requires independent thought, ingenuity and skill, producing in a distinctive form a more efficient result, converting a comparatively defective apparatus into a useful and efficient one, rejecting what is bad and useless in former attempts and retaining what is useful, and uniting them all into an apparatus which, taken as a whole, is novel, there is subject matter. A new combination of well known devices, and the application thereof to a new and useful purpose, may require invention to produce it, and may be good subject-matter for a patent."

Discovery alone that an apparatus, for example, may be altered to produce a new result will not qualify for a patent; one must show that ingenuity has been applied to the discovery to produce a novel and useful method or result.

A patent may not be obtained where an application to patent the same invention has already been filed or the invention has been in public use or disclosed to the public.

Term of Patent

The term of a patent is twenty years from the date of the application for the patent. If a particular invention has been granted a patent, no other valid patent can be granted with respect to that invention.

[6] [1930] A.C. 97

Assignment and Licensing of Patent Rights

The value of a patent is enhanced by virtue of the fact that patent rights can be assigned to others provided the assignment is in writing.

The owner of a patent can assign part or all of the patent rights, in whole or in part, and for such valuable consideration as may be negotiated with the assignee.

Patent rights may also be licensed, on an exclusive or a non-exclusive basis. Usually a royalty fee is charged based on a percentage of sales of the patented product.

Any assignment of a patent right or grant of exclusive licensing rights must be registered in the Patent Office. Otherwise it will be void and therefore unenforceable against a subsequent assignee or exclusive licensee who does register.

Infringement of Patents

The owner of an issued patent has the exclusive right to use or license the patented invention. Infringement of a patent entitles the owner of the patent to claim for all damages sustained — and any damages sustained by the owner's licensees — by reason of such infringement.

To recover damages, a court action may be brought. The court will have to decide whether the patent is valid and the defendant has taken the substance of the plaintiff's invention. If the court decides in favour of the inventor, the inventor may obtain an injunction to restrain the defendant from further infringement. The inventor may then proceed to recover any profits the defendant received as a result of the infringement, based on an accounting thereof. Or the inventor may recover any damages that he or she can prove were incurred as a result of the infringement. To calculate damages, the general approach is to determine the net profit the plaintiff would have made if he or she — rather than the infringing defendant — had sold items produced by the infringement of the patent.

Assignment of Patent Rights by Employee Engineer

Generally it is the inventor who is entitled to apply for a patent. An engineer may be requested, by his or her employer, to execute an agreement that assigns to the employer some or all patent rights to which the engineer might otherwise become entitled during the course of employment.

In general, in the absence of a special contract, the invention of an employee — made in the employer's time, with the employer's materials, and at the expense of the employer — does not become the property of the employer (*Willard's Chocolates Ltd.* v. *Bardsley*[7]). But there is an exception to this general rule if the employee is, by the nature of the employment, expected to apply his or her ingenuity and inventive faculties. Thus if an engineer employed to invent produces a patentable invention during the course of employment, the patent belongs to the employer.

But note that such patent rights extend only to inventions that arise in the course of employment. An excerpt from *British Reinforced Concrete Engineering Co. Limited* v. *Lind*[8] is of interest in this regard:

> In my judgment a draughtsman — I am only dealing with a draughtsman at the moment — in an engineering draughtsman's office, does not do his duty to his employer when he is instructed to prepare a design for the purpose of getting over a known difficulty or for the purpose of arriving at a solution of a problem if he does not exercise among other things such skill and inventive faculty as he may possess. In my judgment it is part of his duty and part of that which he is paid to do, to produce under those circumstances the best design that he is capable of producing. In some cases the employer requires and obtains an express agreement in writing between himself and the employee, providing for inventions that may be made, but in the absence of any agreement of that kind, I desire to guard myself against suggesting that every invention which is made by a person, even though he is a draughtsman in an office in an engineering firm, necessarily belongs to the firm. I am not suggesting that for a moment. I can well conceive that such a person might make an invention in respect of something which is outside his work altogether, having nothing to do with the work upon which his employers are engaged, in which case such an invention would be the draughtsman's own property. But in my judgment, if a draughtsman is instructed by his employer to prepare a design for the purpose of solving a particular difficulty or problem, it is his business and his duty to do the best he can to

---

[7]   [1928–29] 35 O.W.N. 92
[8]   [1917] 34 R.P.C. 101

produce the best design, using all the abilities which he may have, and if he does produce a design which solves the problem, as a result of the instructions that he has been given by his employer, then *prima facie* at any rate the design and then invention are the property of the employer and not of the employee.

That being the law as I understand it, in some cases it may seem to work hardly upon the employee. Taking the present case, the Defendant was a person receiving a comparatively small wage, a man of ability, and I have no doubt of some inventive faculty, and it may seem hard that the fruits of his brains and labours should belong to his employer. On the other hand, the position would be an extraordinarily difficult one if it were not so. The position would be, that a person in the position of a draughtsman, being instructed to prepare a design to get over a known difficulty, and preparing that design and thus solving the difficulty, would be entitled to prevent his employer, who had paid him to do the job and in whose time he had done the work, from making use of that design without coming to some agreement with him, the employee, as to the terms upon which it was to be used. It seems to me that that would be a position which would be almost impossible.

TRADE-MARKS

Definition

"Trade-mark" is defined in the Trade-Marks Act. Part of the definition states:

> "trade-mark" means
> (a) a mark that is used by a person for the purpose of distinguishing wares or services manufactured, sold, leased, hired or performed by him from those manufactured, sold, leased, hired or performed by others.

A trade-mark may be registered according to the Trade-Marks Act in connection with wares and/or services. Unless shown to be invalid, the registration of a trade-mark gives the owner the exclusive right to use the trade-mark throughout Canada in association with the wares or services covered by the registration.

An essential feature of a valid trade-mark is its distinctiveness. The trade-mark must distinguish goods of one manufacturer from those of another; manufacturers must not deceive the public. The Trade-Marks Act provides:

> "distinctive" in relation to a trade-mark means a trade-mark that actually distinguishes the wares or services in association with which it is used by its owner from the wares or services of others or is adapted so to distinguish them....

A trade-mark registration that ceases to be distinctive may be cancelled.

When Trade-Marks Are Registrable

The Trade-Marks Act lists certain words and marks that may be registered as trade-marks. Section 12(1) provides, in part:

> Subject to section 13, a trade-mark is registrable if it is not
> (a) a word that is primarily merely the name or the surname of an individual who is living or has died within the preceding thirty years;
> (b) whether depicted, written or sounded, either clearly descriptive or deceptively misdescriptive in the English or French languages of the character or quality of the wares or services in association with which it is used or proposed to be used or of the conditions of or the persons employed in their production or of their place of origin;
> (c) the name in any language of any of the wares or services in connection with which it is used or proposed to be used;
> (d) confusing with a registered trade-mark.

Licensing

The Trade-Marks Act allows a trade-mark to be used by third parties. In order to maintain the distinctiveness of a licensed mark, the third parties must be licensed under the authority of the trade-mark owner. The terms of the licence agreement must provide the trade-mark owner with direct or indirect control of the character or quality of the wares or services to be provided by the licensee. In addition, if a trade-mark is used by a licensee, it is important to include in packaging and advertising materials a notice of the identity of the trade-mark owner and the existence of the licence.

Duration of Registration

The Trade-Marks Act prescribes that registrations are effective for a period of fifteen years; registrations may be renewed for unlimited subsequent periods of fifteen years each. A trade-mark registration may be cancelled if the owner stops using the trade-mark.

Infringement

A person who infringes a valid registered trade-mark by using the same or a confusing mark may be restrained from continuing to use the mark; that person may also be liable for damages that resulted from the infringing of the trade-mark. Forgery of a trade-mark with intent to deceive or defraud the public or any person is an offence under the Criminal Code of Canada. The offence is punishable by fine and imprisonment for up to two years.

PASSING OFF

The owner of an unregistered trade-mark may sue a defendant who uses the same or a similar mark in passing off. To succeed, the plaintiff must show the mark is identified with the plaintiff's wares or service in the relevant marketplace and that the defendant's use of the same or a similar mark is causing customers to purchase from the defendant in the mistaken belief that they are dealing with the plaintiff. The remedies for passing off are the same as are available to the owner of a registered mark whose mark is infringed.

COPYRIGHT

Definition

Subject to the provisions of the Copyright Act, copyright subsists in every original literary, dramatic, musical, and artistic work. "Copyright" generally means the sole right to produce or reproduce the work, or any substantial part thereof in any material form whatever. As well, it conveys sole right to perform — and, in the case of a lecture, to deliver — the work or any substantial part thereof in public. If the work is unpublished, copyright conveys sole right to publish the work, or any substantial part thereof.

Copyright does not protect designs applied to useful articles that are mass produced.

Term of Copyright

Except as otherwise expressly provided by the Act, copyright subsists for a term that equals the life of the author and a period of fifty years after the author's death.

Registration of Copyright

The Copyright Act provides that an author, the author's legal representatives, or an agent may apply for the registration of a copyright at the Copyright Office. The particulars of any assignment or licence of a copyright may also be registered at the Copyright Office.

Registration is not essential to copyright. However, registration of copyright in a work may assist the owners in obtaining damages for infringement. Additionally, registration of assignments and licences of copyright are advisable in order to protect assignees and licensees from claims of other subsequent assignees and licensees who may register.

Ownership of Copyright

The first owner of copyright in a work is the author of the work. The owner of the copyright is entitled to assign the copyright in whole or in part. An assignment may be made subject to territorial and timing limitations. The assignment must be in writing to be effective.

Moral Rights

The creator of a work in which copyright subsists also has moral rights in the work. Moral rights include the right to be identified as the author of the work by name or pseudonym, or the right to remain anonymous, and the right to the integrity of the work. The right to integrity is infringed if the work is altered in any manner or used in association with a product, service, cause, or institution and such alteration or use affects the honour or reputation of the author. The remedies for infringement of moral rights are the same as the remedies for copyright infringement. Moral rights may be waived but cannot be assigned. The waiver must be in writing.

Engineering Plans

Copyright in engineering plans generally belongs to the engineer who authored the plans or to the engineer's employer if the plans were created in the course of such employment. Where an engineer prepares the plans for a client, and unless otherwise agreed, the client is precluded from reproducing the engineer's plans or repeating his or her design in a new structure without the express or implied consent of the engineer. Engineer-client agreements should include a provision that draws attention to this aspect of the engineer's copyright.

INDUSTRIAL DESIGNS

Definition

The Industrial Design Act grants protection to originators of certain industrial designs. The term "industrial design" refers to any features of shape, configuration, pattern, or ornament that are applied to finished articles and appeal to and are judged solely by the eye where the articles are multiplied by an industrial process. Only those designs that are ornamental or aesthetic in nature — as opposed to functional — qualify for protection under the Industrial Design Act. The mechanical construction of an article does not form part of the design; neither does the method of manufacture. Construction and method are functional, and thus are not protected by the Act. (Mechanical constructions or methods of manufacture may qualify for patent protection.)

In order to qualify for protection, a design must meet the tests of novelty and originality, and an application to register the design must be filed within one year of publication.

Term

The proprietor who registers an industrial design is granted an exclusive right to the use of the design for a term of five years. The term is subject to renewal for an additional five years.

Assignment

Designs are freely assignable; the assignment must be made in writing. In addition, a proprietor may license others to make, use, or sell the design during the term of its statutory protection.

Employees

The rights to any designs made by an employee in the course of employment belong to the employer.

Registration

To register a design under the Industrial Design Act, a proprietor submits a drawing and description of the design in duplicate, together with the prescribed fee, to the appropriate government office. Registration will be refused if it appears that the design is identical with or closely resembles another design currently in use or previously registered.

TRADE SECRETS

A patent of invention provides monopoly rights for a limited period; that is, for the term of the patent. Once an invention has been patented, its subject matter is no longer private. Full details of the invention are publicized and become part of the "public domain." When the term of the patent expires, the invention may be freely used by others. The limited term of patent protection causes some concern; as well, it is sometimes difficult to enforce a patent. To avoid these problems, inventors sometimes do not obtain patent protection but rely instead on trade secret protection.

The Nature of a Trade Secret

The subject matter of "trade secrets" or "confidential information" has been defined by the courts[9] as follows:

> A trade secret may consist of a formula, pattern, device or compilation of information which is used in one's business and which gives him an opportunity to obtain an advantage over competitors who do not know or use it. It may be a

---

[9] Seager v. Copydex Ltd. [1967] R.P.C. 349

formula for a chemical compound, a process for manufacturing, treating or preserving materials, a pattern for a machine or other device or a list of customers.

The definition includes what is generally referred to as "industrial know-how." The term describes valuable information acquired by a business enterprise, for example, marketing and manufacturing techniques, organizational methods, and technical data.

Patent law does not effectively protect much of the subject matter of trade secrets. In fact, most elements of "know-how" — such as organizational methods and customer lists — are unpatentable. Trade-secret protection is the only available legal remedy for the unauthorized use of such information.

A person who wishes to succeed in an action for unauthorized disclosure or use of a trade secret must establish two elements. First, it must be shown that the information was indeed "know-how" — the necessary quality of secrecy must be adduced. Second, it must be shown that the secret information was communicated to the defendant in circumstances that implied a duty of confidence.

Only confidential information — not general knowledge — will qualify for trade-secret protection. If the possessor of a secret voluntarily discloses that secret without restrictions or otherwise fails to take reasonable steps to ensure the secrecy of the information, the protection will be lost. In general, disclosure of the secret must be strictly confined to a limited group of persons, such as employees and potential licensees, or the confidential nature of the secret is lost.

To determine whether certain information possesses the requisite degree of secrecy, the courts have looked at the following factors:

(1) the extent to which the information is known outside the business,

(2) the extent of measures taken to guard the secrecy of the information,

(3) the value of the information to competitors,

(4) the amount of effort or money spent to develop the information.

Employees

The common law recognizes the right of an employer to restrain a former employee from making improper use of trade secrets, and the employer's right to damages for profits resulting from any such improper use.

In *Amber Size & Chemical Co., Ltd.* v. *Menzel*,[10] the court made several remarks that illustrate the principles a court might apply:

> In my view, after giving the authorities the best attention I can, the law stands thus: — The Court will restrain an ex-servant from publishing or divulging that which has been communicated to him in confidence or under a contract by him, express or implied, not to do so.... and generally from making an improper use of information obtained in the course of confidential employment.... and, further, from using to his late master's detriment information and knowledge surreptitiously obtained from him during his, the servant's employment....
>
> In applying these principles I have to answer four questions of fact:
>
> First, did the plaintiffs in fact possess and exercise a secret process?
>
> Secondly, did the defendant during the course of his employment know that this process was secret?
>
> Thirdly, did the defendant acquire knowledge during his employment of that secret or a material part thereof?
>
> Fourthly, has he since leaving the plaintiffs' employ made an improper use of the knowledge so acquired by him?

Duty of Confidence   In the relationship of employer and employee, it is an implied contractual term that the employee will not disclose the confidential information of the employer. This obligation continues even after the employment is terminated. But the employee may use any ordinary working knowledge and general experience acquired in one job in any subsequent employment. It is often difficult to distinguish between ordinary working knowledge and protectable trade secrets.

In general, the courts have demonstrated an extreme reluctance to prohibit an employee from using his or her skills in subsequent employment. Clear and convincing evidence of confidentiality is required when an employer alleges breach of confidence by a former employee.

[10] [1913] 2 Ch. 239

An obligation of confidence also exists where the owner of a trade secret discloses the secret to another enterprise for a specific purpose, such as having a mould or dye made for future production.[11]

In the process of negotiating for the licensed use by the recipient of a trade secret, the relationship between two parties will give rise to an obligation of confidence even if no agreement on the use of the trade secret is ever entered into.[12]

A party that uses or discloses a trade secret in breach of a duty of confidence may be restrained from using the trade secret and may be liable to pay the owner damages or the profits earned by that party from using the trade secret. Third parties who learn the trade secret from a party who has breached a duty of confidence may have similar liability.

[11] Saltman v. Campbell [1948] 65 R.P.C. 203
[12] Coco v. Clark [1969] R.P.C. 41

CHAPTER THIRTY-FOUR

THE LAW OF QUEBEC

INTRODUCTION

Engineers involved in construction projects in the Province of Quebec should take note that there are certain fundamental differences between the laws applicable in the Province of Quebec and the laws applicable in the common-law provinces. Numerous statutes adopted by the National Assembly of Quebec contain specific provisions that apply to construction projects. These statutes govern such matters as the professional qualification of contractors, the qualification of construction workers, health and safety on the work site, protection of the environment, and safety requirements for public buildings.

The Engineers' Act of Quebec, R.S.Q. c.I-9 enumerates the services that constitute the exclusive field of practice of an engineer and sets out the conditions under which persons who are not domiciled in Quebec can either be admitted as members of the Order of Engineers of Quebec or can obtain a temporary licence for a specific project.

Many provisions of law that govern the performance of engineering services in the Province of Quebec, including such matters as the liability of engineers to their clients and to third parties, are contained in the Civil Code of Quebec.

THE CIVIL CODE OF QUEBEC

The Civil Code of Quebec came into force on January 1, 1994. It replaced the Civil Code of Lower Canada which had originally been adopted in 1865. The Civil Code of Quebec is considered the cornerstone of the Quebec civil law system. This is confirmed by the preliminary provision of the Civil Code:

The Civil Code of Québec, in harmony with the Charter of human rights and freedoms and the general principles of law, governs persons, relations between persons, and property.

The Civil Code comprises a body of rules which, in all matters within the letter, spirit or object of its provisions, lays down the jus commune, expressly or by implication. In these matters, the Code is the foundation of all other laws, although other laws may complement the Code or make exceptions to it.

In order to protect their interests and the interests of their clients when doing business in the Province of Quebec, engineers should have a basic understanding of the provisions of the Civil Code that are likely to apply to the conclusion and execution of construction contracts.

CONTRACTUAL AND EXTRACONTRACTUAL LIABILITY

Under Quebec law, liability exists when there is a breach of an obligation. An obligation can arise from a contract or from any act or fact to which the effects of an obligation are attached by law. An obligation that arises from a contract is a contractual obligation, whereas an obligation that arises from the law is referred to as an extracontractual obligation.

The Civil Code provides that every person has a duty to abide by the rules of conduct that apply to him or her according to the circumstances, usage, or law so as not to cause injury to another person. When a person fails to respect this duty, he or she is responsible for any bodily, moral, or material injury he or she causes to another person.

The courts will exercise their discretion in deciding whether or not a particular act or omission constitutes a breach of an obligation. In professional liability cases, the courts will consider how a normally prudent and competent member of the same profession would have acted under the same circumstances.

The Civil Code also states that a person may be held liable for the injuries caused to another by the act or fault of another person. For example, an employer can be held responsible for the injuries or damages caused by its employee. Furthermore, there is a presumption of liability against a person for damages caused by property in his or her custody.

INTERPRETATION OF CONTRACTS

The Civil Code contains numerous provisions dealing with the interpretation of contracts. Certain provisions reflect a desire to protect the consumer and the party who signs a "contract of adhesion." A contract of adhesion is one where the terms are drafted by one party without the other party having the opportunity of negotiating or changing them.

The classification of a contract as a consumer contract or a contract of adhesion will have an impact on the interpretation of the effects of the contract. For example, in the case of a consumer contract or a contract of adhesion, an external clause referred to in the contract will be deemed null if at the time of the conclusion of the contract it was not brought expressly to the attention of the consumer or the adhering party, unless the other party proves that the consumer or adhering party otherwise knew of it.

Engineers who are involved in the drafting of contracts for construction projects in the Province of Quebec should be aware of this provision and should ensure that all documents and clauses to which the contract refers form part of the contract documents. If this is not practical, then the consumer or adhering party should acknowledge in writing that the external clause referred to in the contract was expressly brought to his or her attention.

The Civil Code also states that in a consumer contract or contract of adhesion, a clause that is illegible or incomprehensible to a reasonable person is null if the consumer or the adhering party suffers a prejudice as a result thereof. Evidence that an adequate explanation of the nature and scope of the clause was given to the consumer or adhering party would, however, constitute a valid defence to such a claim.

In a consumer contract or contract of adhesion, a clause that is excessive or unreasonably detrimental to the consumer or the adhering party may be considered abusive, in which case it can be declared null or the obligation arising from it may be reduced.

These provisions of the Civil Code give the courts the power to reduce the contractual obligations of a party in certain circumstances. Care must therefore be taken in the drafting of contracts, to avoid the inclusion of clauses that may be considered by a court to be illegible, incomprehensible, or abusive.

LIMITATION OF LIABILITY

A person may not exclude or limit his or her liability for material damages caused to another through an intentional fault or gross negligence, nor is it possible to exclude or limit one's liability for bodily or moral injury.

A notice stipulating the exclusion or limitation of the obligation to repair an injury resulting from the non-performance of a contractual obligation is only valid if the party invoking the notice proves that the other party was aware of its existence at the time the contract was formed.

A person may not by way of a notice limit or exclude his or her obligation to repair the injury caused to a third person, but such a notice may constitute a warning of danger.

FORCE MAJEURE

The Civil Code provides that a person may exculpate himself or herself from liability for injury caused to another by proving that the injury results from an unforseeable and irresistible event. The appreciation of what constitutes a case of *force majeure* is left to the discretion of the courts.

DISCLOSURE OF TRADE SECRETS

A person may exculpate himself or herself from liability for injury caused to another as a result of the disclosure of a trade secret by proving that considerations of general interest prevail over keeping the secret and, particularly, that its disclosure was justified for reasons of public health or safety. Although this provision may not necessarily receive application in construction matters, it does confirm the importance that the Civil Code attaches to the health and safety of the public.

SOLIDARITY

Where several persons have jointly taken part in a wrongful act that has resulted in an injury, or if they committed separate faults each of which may have caused the injury, and where it is impossible to determine which of them actually caused it, they are solidarily (jointly and severally) liable for reparation thereof.

An obligation is solidary (joint and several) between debtors where they are obligated to the creditor for the same thing in such a way that each of them may be compelled separately to perform the entire obligation and where performance by a single debtor releases the others towards the creditor.

Solidarity between debtors exists where it is expressly stipulated by the parties or imposed by law. It is presumed to exist where an obligation is contracted for the performance of a service or an organized economic activity. Engineers who participate in the execution of construction projects as members of a joint venture should therefore consider to what extent they may be held solidarily liable for the faults of the other members of the joint venture.

The obligation to repair the injury caused to another through the fault of two or more persons is solidary where the obligation is extracontractual.

REMEDY FOR NON-PERFORMANCE

An obligation gives the creditor the right to demand that it be performed in full, properly, and without delay.

When the debtor fails to perform his or her obligation without justification, the creditor may either enforce specific performance of the obligation or obtain, in the case of a contractual obligation, the resolution or resiliation of the contract. The creditor can also ask for the reduction of the creditor's own obligations.

If the debtor fails to respect his or her obligations, the creditor may perform the obligation or cause it to be performed at the expense of the debtor. A creditor wishing to do so must first put the debtor in default, except in cases where the debtor is in default by operation of the law or by the terms of the contract.

This would apply, for example, in the case where a contractor fails to properly execute a construction contract. After having put the contractor in default, the owner can have the work completed by another contractor and claim any additional cost from the defaulting contractor.

ASSESSMENT OF DAMAGES

A creditor is entitled to damages for bodily, moral, or material injuries that are an immediate and direct consequence of the debtor's default. The obligation of the debtor to pay damages to the creditor is neither reduced nor altered by the fact that the creditor receives compensation from a third person as a result of the injury the creditor has sustained, except if the third party is subrogated in the rights of the creditor. An example of such a case is where the creditor who suffers a loss is indemnified by his or her insurance company.

The damages awarded to the creditor are intended to compensate the amount of the loss the creditor has sustained and the profit of which the creditor has been deprived. Future loss that is certain and can be assessed is taken into account in awarding damages. For example, if a contractor's negligence results in the late completion of the construction of an apartment building, the owner can claim compensation for loss of revenue if the owner can prove that the apartments would have been rented immediately upon completion of the work.

In contractual matters, the debtor is only liable for damages that were foreseen or foreseeable at the time the obligation was contracted unless the non-performance of the obligation is the result of an intentional fault or gross negligence. Even then, only damages that are an immediate and direct consequence of the non-performance will be awarded.

PENAL CLAUSES

A penal clause is a clause in virtue of which the parties assess the anticipated damages by stipulating that the debtor will pay a penalty if the debtor fails to perform the debtor's obligation. The creditor can avail himself or herself of the penal clause instead of enforcing the specific performance of the obligation, but in no case may the creditor claim both the performance of the obligation and the penalty unless the penalty has been stipulated for a mere delay in the performance of the obligation. For example, if a contractor is late in the completion of the construction of a building and the contract contains a penalty clause for late delivery, the owner can require the contractor to complete the construction and the owner may in addition claim the penalty for late delivery.

MANUFACTURER'S LIABILITY

The manufacturer of a product is liable for the injury caused to a third person by reason of a safety defect in the product. The same rule applies to a person who distributes the product under his or her name or as his or her own product, and to any supplier of the product. A product is deemed to have a safety defect if it does not afford the safety that a person is normally entitled to expect, by reason of a defect in the design or manufacture of the product, the preservation or the presentation of the product, or the lack of sufficient indications as to the risks, dangers, or safety precautions.

The seller of a product must warrant to the buyer that the product and its accessories are, at the time of sale, free of latent defects that render it unfit for the use for which it was intended or that so diminish its usefulness that the buyer would have not bought it, or paid such a high price had the buyer been aware of the defects. The seller, however, is not bound to warrant against any apparent defect or latent defect known to the buyer.

A defect is presumed to have existed at the time of a sale by a professional seller if the product malfunctions or deteriorates prematurely in comparison with products of the same type. Such a presumption does not apply if the defect is due to improper use of the property by the buyer.

The manufacturer, any person who distributes the property under his or her name or as his or her own, and any supplier of the property, in particular the wholesaler and the importer, are also bound to warrant the buyer in the same manner as the seller.

THE CONTRACT OF ENTERPRISE OR FOR SERVICES

The contract of enterprise or for services is a contract by which a person, referred to as the contractor, or the provider of services (such as an engineer), undertakes to carry out physical or intellectual work for a client, or undertakes to provide a service for a price that the client agrees to pay.

One of the major differences between the contract of enterprise or for services and the contract of employment is that the contractor, or the provider of services, is free to choose the method of performance of the contract and no relationship of subordination exists between the contractor, or the provider of services, and the client in respect of such performance.

The contractor and the provider of services must act with prudence and diligence and in the best interest of their client. Depending on the nature of the work, they are bound to act in accordance with the practice of the trade, the rules of the art, and the terms of the contract.

Before the contract is entered into, the contractor or the provider of services is bound to provide to the client, as far as the circumstances permit, any useful information concerning the nature of the work that they undertake to perform and the property and time required for the task.

The Civil Code imposes on the engineer the obligation to furnish all relevant information to the client concerning the performance of the contract even if the client does not ask for it. The determination of the information that must be furnished is left to the discretion of the courts and depends on the circumstances of each case.

When the price of the work or services is fixed by the contract, the client must pay this price and the client cannot claim a reduction on the grounds that the work or services cost less than anticipated.

On the other hand, in the case of a fixed price contract, the contractor and the provider of services may not claim an increase in the price of the contract even if the work or services cost more than had been anticipated.

In the case where the price of the work is estimated at the time the contract is entered into, the contractor or provider of services must justify any increase in the price.

PRESUMPTION OF LIABILITY

The provisions of the Civil Code cited below establish a presumption of liability applicable to the parties who participate in the construction of a building; these provisions of the Civil Code are considered to be of public order and may not be derogated from, even by private agreement:

> 2118. Unless they can be relieved from liability, the contractor, the architect and the engineer who, as the case may be, directed or supervised the work, and the subcontractor with respect to work performed by him, are solidarily liable for the loss of the work occurring within five years after the work was completed, whether the loss results from faulty design, construction or production of the work, or the unfavourable nature of the ground.

> 2119. The architect or the engineer may be relieved from liability only by proving that the defects in the work or in the

part of it completed do not result from any erroneous or faulty expert opinion or plan he may have submitted or from any failure to direct or supervise the work.

The contractor may be relieved from liability only by proving that the defects result from an erroneous or faulty expert opinion or plan of the architect or engineer selected by the client. The subcontractor may be relieved from liability only by proving that the defects result from decisions made by the contractor or from the expert opinion or plans furnished by the architect or engineer.

They may, in addition, be relieved from liability by proving that the defects result from decisions imposed by the client in selecting the land or materials, or the subcontractors, experts, or construction methods.

The owner of a building that perishes in whole or in part within five years after completion can therefore rely on this presumption of liability of the architect, engineer, contractor, and subcontractors without having to prove the precise cause of the loss. In order to rebut the presumption of liability, the architect, engineer, contractor, and subcontractor have the burden of proving that they committed no fault, or that the defects result from a decision imposed by the client.

Article 2120 of the Civil Code stipulates that the contractor, the architect, and the engineer, in respect of work they directed or supervised, and where applicable, the subcontractor in respect of work the subcontractor performed, are *jointly* liable to the owner for defects due to poor workmanship, which exist at the time of acceptance of the work, or which are discovered within one year after acceptance.

It is to be noted that there is a difference between the French and the English text of article 2120 of the Civil Code. The French text mentions that the contractor, architect, engineer, contractor, and subcontractor are *jointly* liable. The English text states that they are *solidarily* liable. If they are jointly liable, they are only responsible to the owner for their proportionate share of the cost of the repairs. For example, if an architect, an engineer, and a contractor are involved, each would be responsible for one-third of the cost. If they are solidarily liable, each party is liable to the owner for the total cost of the repairs, subject to that party's right to claim reimbursement of the share due by each of the other two parties.

The comments of the Ministry of Justice concerning the interpretation of the articles of the Civil Code confirm that the French version is correct and that the contractor, architect, engineer, and subcontractors are *jointly* and not *solidarily* responsible to the owner for defects due to poor workmanship.

The architect or the engineer who does not direct or supervise the work is only liable for the loss occasioned by a defect or error in the plans or the expert opinions furnished by the architect or engineer.

LEGAL HYPOTHECS

A hypothec is a real right affecting movable or immovable property. It gives the creditor the right to exercise his or her rights against the property even if it has changed hands. It also gives the creditor the right to take possession of the property, to take it in payment, to sell it, or cause it to be sold, and in such a case to receive the proceeds of the sale by preference, according to the rank determined in the Civil Code.

A legal hypothec in favour of persons who take part in the construction or renovation of an immovable (building) can only affect that immovable. It exists in favour of the architect, engineer, supplier of materials, workers, and contractor or subcontractors for the work requested by the owner of the immovable and for the materials or services supplied or prepared for such work.

The legal hypothec in favour of persons taking part in the construction or renovation of an immovable subsists for a period of 30 days after the work has been completed. It continues to subsist if, before the 30-day period expires, a notice describing the affected immovable and indicating the amount of the claim is registered in accordance with the relevant formalities. The notice must be served on the owner of the immovable. The legal hypothec will become extinguished six months after the work is completed unless the creditor publishes an action against the owner of the immovable or registers a prior notice of exercise of the creditor's hypothecary right.

The hypothec secures the increase in the value added to the immovable by the work, materials, or services supplied or prepared for the project. However, where the persons in whose favour it exists did not enter into a contract directly with the owner, the hypothec is limited to the work, materials, or services supplied after written notification of the contract to the owner. Subcontractors and suppliers of materials who enter into a contract with the general contractor must therefore give a written notice to the owner of the immovable before commencing the work in order to preserve their right to register a legal hypothec. Workers are not required to give such a notice.

At the request of the owner of the property affected by a legal hypothec, the courts may authorize the substitution of another form of security in order to guarantee payment to the creditor.

Engineers who administer construction contracts for their clients must take the necessary precautions in order to ensure that the parties who have the right to register legal hypothecs are paid for their work and services. Otherwise, these creditors may register legal hypothecs against the property and oblige the owner to satisfy their claims. If the contractor is insolvent, the owner may never be able to recover these amounts from him or her.

The right to register a legal hypothec also benefits the engineer if the engineer's fees for professional services are not paid.

CONCLUSIONS

It is impossible to summarize all the important provisions of law applicable to construction projects in Quebec. Accordingly, comments have been provided on selected aspects of the law contained in the Civil Code that are most likely to be of interest to an engineer involved in a project in the Province of Quebec.

Certain provisions of law contained in the Civil Code are new and original. Since they have not yet been judicially interpreted, the decisions rendered by the courts in the years to come will be of significant importance.

It is wise to seek legal advice from an expert in the field as soon as a litigious situation arises. It is even wiser to seek legal advice at the contract review stage, since many problems of interpretation and many claims can be avoided by the use of properly drafted contract documents.

As indicated in the preface to this text, this chapter (dated August 15, 1995) was kindly prepared by Olivier F. Kott, a Quebec lawyer and partner of Ogilvy Renault.

CHAPTER THIRTY-FIVE

THE NORTH AMERICAN
FREE TRADE AGREEMENT

Although not as widely understood as it should be, Canada and the U.S. are the world's two largest trading partners. In the first ten months following the implementation of the North American Free Trade Agreement ("NAFTA"), Canadian exports to the U.S. rose 21.1 percent from the previous year to a total of $144.9 billion. During the same period, American exports to Canada rose 19.3 percent to $123.1 billion. Imports from Mexico into Canada increased 23 percent to $3.64 billion.

There are strong similarities in the approach to business in Canada and the United States.

NAFTA should provide new opportunities for Canadian engineers and contractors. An understanding of NAFTA is important to those who may seek out those opportunities.

THE CANADA–UNITED STATES FREE TRADE AGREEMENT

Prior to free trade also extending to Mexico under NAFTA in 1994, the Canada–United States Free Trade Agreement (the "FTA") became effective on January 1, 1989.

National Treatment

Under the FTA, Canada and the United States agreed to provide "national treatment" to investors from each other's country in relation to the establishment of new businesses, the acquisition of existing businesses (subject to certain monetary thresholds) pursuant to the Investment Canada Act, and the conduct, operation, and sale of established businesses.

As an exception to the national treatment principle, the FTA provided that each country may accord investors of the other country different treatment provided that:

- the difference in treatment is no greater than that necessary for prudential, fiduciary, health and safety, or consumer protection reasons;

- such different treatment is equivalent in effect to the treatment accorded by the country to its own investors for such reasons; and

- prior notification of the proposed treatment is given in accordance with the terms of the FTA.

Prohibition on Performance Requirements

The FTA also provides that the parties are not to impose on investors from the other country, as a condition of permitting an investment or regulating a business, requirements relating to any minimum volume of exports of goods or services, the achievement of minimum amounts of domestic content, local sourcing, or import substitution. Additionally, the FTA provides that neither country shall impose any of the foregoing performance requirements on an investor from a third country where meeting the requirement could have a significant impact on trade between the two countries.

No Minimum National Equity Ownership Requirements

The FTA provides that neither country will adopt policies requiring minimum levels of equity (ownership) holdings by their nationals in domestic firms controlled by investors from the other country or requiring forced divestiture (subject to an exception for cultural industries).

Expropriation

Neither country is permitted directly or indirectly to nationalize or expropriate an investment in its territory by an investor of the other country except for a public purpose, in accordance with due process of law, on a non-discriminatory basis, and upon payment of prompt, adequate, and effective compensation at fair market value. The FTA also provides for the free transfer of profits and other remittances subject only to certain exceptions relating to bankruptcy, criminal offences, reports of currency transfers, withholding taxes, trading or dealing in securities, or ensuring the satisfaction of judgments.

Current Status of FTA

However, the passage of NAFTA has resulted, in effect, in the suspension of the FTA. Technically, it remains in place, but is not of any real effect while NAFTA is in place. However, certain important aspects of the FTA, such as tariff elimination between the U.S. and Canada, remain operative.

NORTH AMERICAN FREE TRADE AGREEMENT

On January 1, 1994, NAFTA — between Canada, Mexico, and the United States — came into effect. The investment provisions of NAFTA are closely modelled on the counterpart provisions of the FTA.

Additional Provisions

There are certain additional provisions that are included in NAFTA but are not currently contained in the FTA. These include the principles of "most-favoured-nation treatment" — obligating each NAFTA country to accord investors of each other NAFTA country treatment no less favourable than it accords in like circumstances to investors of another NAFTA country or of a non-NAFTA country. In addition, each NAFTA country is to accord investments of investors of each other NAFTA country treatment in accordance with international law, including fair and equitable treatment and full protection and security.

Performance Requirements

NAFTA adds to those proscribed under the FTA prohibitions with respect to performance requirements that tie the volume or value of imports or sales within a territory to the volume or value of exports from the host country or to its foreign exchange earnings, as well as provisions relating to technology transfers and product mandating.

Similarly, NAFTA prohibits any NAFTA country from imposing conditions on investors of another NAFTA country for preferential sourcing of goods in the host country or minimum host country content.

NAFTA prohibits a NAFTA country from requiring that an enterprise of another NAFTA country appoint to senior management positions individuals of any particular nationality. However, a NAFTA country may require that a majority of the board of directors or any committee of the board of such an enterprise must be of a particular nationality or resident in the territory of the host NAFTA country, so long as that requirement does not materially impair the ability of the investor to exercise control over its investment.

Exceptions

NAFTA's principles respecting "national treatment," "most-favoured-nation treatment," performance requirements, and senior management and boards of directors do not apply to any existing non-conforming measure (such as the Investment Canada Act).

The national treatment, most-favoured-nation treatment, and senior management principles are, however, not applicable to government procurement of goods or services or subsidies and grants provided to a NAFTA country, including government-support loans, guarantees, and insurance. This means that NAFTA countries may discriminate against the investors of other NAFTA countries, or non-NAFTA countries for that matter, in regard to government procurement of goods and services or in the granting of subsidies or grants.

Investor Dispute Resolution

NAFTA sets out a comprehensive code for the resolution of investment disputes involving a breach or an alleged breach of NAFTA investment rules by a NAFTA country. A NAFTA investor may either seek monetary damages through binding investor-state arbitration or remedies that are available in the domestic courts of the host country. The FTA contained no provisions specifically enabling investors to require the resolution of investment disputes with a host country directly. This situation was remedied in the investor dispute resolution provisions contained in NAFTA.

CONSTRUCTION AND NAFTA

Cross-Border Trade in Services and NAFTA

In extending free trade commitments to construction services, NAFTA breaks significant new ground. NAFTA opens the construction markets of each country to construction service providers from the other NAFTA countries. Accordingly, it prohibits American contractors in Canada from receiving less favourable treatment than Canadian contractors, and vice versa. For example, American contractors are not required to incorporate in Canada in order to provide services north of the border. NAFTA also prohibits any requirement that American firms have an office or a residence in Canada as a condition of providing construction services in Canada.

As a result, opportunities for Canadian engineers in the U.S. and in Mexico should expand. To ensure that licensing and certification of nationals (engineers, for example) of another NAFTA country do not constitute unnecessary barriers to trade, NAFTA provides that such measures must:

• be based on objective and transparent criteria such as professional competence;

• not be more burdensome than necessary to ensure the quality of the service; and

• not constitute a restriction on the cross-border provision of a service.

The Canadian Council of Professional Engineers ("CCPE") has announced the signing of an important agreement by Canada, the U.S., and Mexico that enables appropriately experienced Canadian engineers to practise in those other jurisdictions, without having to write examinations. In making the announcement, the president of the CCPE pointed out that engineers should check with their respective provincial engineering associations before embarking on business initiatives in the other jurisdictions, as ratification of the agreement by each provincial association is required. The engineer should also check with the regulatory authority that governs the practice of engineering in the other jurisdiction where the engineer intends to practise.

The signing of this agreement is indeed a significant achievement. One that should assist the engineering profession in opening new markets and in implementing the cross-border performance of professional services in Canada, the U.S., and Mexico as contemplated by NAFTA.

Procurement Under NAFTA

NAFTA is the first international trading agreement that applies to construction services. NAFTA's government procurement provisions apply to service contracting and construction in addition to goods. Procurement procedures under NAFTA are designed to ensure that each of the three parties' procedures are fair and equal and entitle open access to each party's procurement opportunities.

Apart from the FTA and NAFTA, a Canadian Content Policy applies generally to tenders for goods valued at between Cdn. $25,000 and $31,800 and to tenders for services between Cdn. $25,000 and Cdn. $60,000. Tenders for goods over Cdn. $31,800 and for services in excess of Cdn. $60,000 take place under the Open Bidding System and the Canadian Content Policy does not apply. Tenders on goods or services valued under Cdn. $25,000 are subject to yet a third procedure whereby the Canadian Content Policy does not apply. However, tenders for goods or services under Cdn. $25,000 are conducted using source lists, rather than on the basis of the open system. Under this procedure, any Department of Supply and Services qualified supplier on the source list is free to participate in the process. These threshold amounts may change over time.

The threshold values of contracts subject to NAFTA depend on the customer and the subject matter of the procurement. For example, for Canadian federal government "entities" (of which 100 are listed in NAFTA) the threshold value for contracts of goods, services, or any combination thereof is U.S. $50,000, and U.S. $6.5 million for contracts of construction services. For government "enterprises" the threshold values are U.S. $250,000 for contracts for goods, services, or a combination thereof and U.S. $8 million for contracts for construction services. The applicable thresholds for state and provincial governments remain to be negotiated.

The NAFTA threshold values are subject to adjustment every second year, starting in 1996, using the indexing and currency conversions in the agreement. The NAFTA parties have agreed to exclude various types of procurement. Canada has exempted shipbuilding and repair, urban rail and urban transportation equipment, systems, components, and materials incorporated therein, as well as project related materials of iron or steel and contracts respecting communications, detection, and coherent radiation equipment. Both Mexico and the United States have excluded procurement related to transportation services. Other types of services excluded include non-contractual agreements and any form of government assistance — including cooperative agreements, grants, loans, equity infusions, guarantees, and fiscal incentives.

Procurement procedures under NAFTA are designed to ensure each of the three parties' procedures are fair and equal and entitle open access to each party's procurement opportunities. NAFTA provisions prohibiting "offsets" require that the parties do not consider, seek, or impose offsets in both the qualification and the selection of suppliers, goods, or services. Offsets are defined as conditions imposed or considered prior to or during an entity's procurement process to encourage local development or improve balance of payment accounts by means of local content requirements, licensing of technology, investment, counter-trade, or similar requirements.

NAFTA also contains mandatory provisions governing tender documents and procedures. Three types of bid procedures are contemplated; open, selective, and limited. The provisions applicable to open bids require that parties issue notices of proposed procurement in the form of invitations to participate that describe: the nature and quantity of goods or services procured; whether the procedure is open or selective and whether it involves negotiation, delivery dates, the contacts for submissions of bids, or the qualifications to be a supplier; the deadline for bid receipt; the identity and address of the entity awarding the contract; a statement of any economic or technical requirements; and the financial guarantees or other documents required. Canada has agreed to publish these notices in *Government Business Opportunities* and on the Open Bidding Service electronic bulletin board.

NAFTA establishes minimum requirements for the use of selective tendering procedures, requiring that parties invite tenders from the maximum number of domestic and other-party suppliers and, upon request, provide reasons for not inviting a tender from a foreign party supplier.

Limited tendering procedures are permitted in the absence of tenders in response to open or selective calls for tenders, or where the bids received result from collusion, where the bids fail to conform with the essential tender documentation requirements, where copyright or patent-related reasons so require, and in matters of urgency.

The construction industry is one of the sectors most affected by NAFTA. NAFTA's provisions should substantially increase the ability of Canadian and American contractors to provide construction services in each other's country without discriminatory restrictions.

TEMPORARY ENTRY OF INDIVIDUALS INTO CANADA UNDER NAFTA

Under the Immigration Act of Canada, only Canadian citizens or permanent residents may work in Canada without a valid employment authorization. As a general rule, a person who is neither a citizen nor a permanent resident of Canada will be authorized to take employment in Canada only if there is no qualified Canadian available to fill the position in question.

The provisions of NAFTA dealing with the temporary entry of certain types of businesspersons are designed to facilitate entry into and travel between NAFTA countries as a necessary complement to the objectives of free trade. Specifically, NAFTA provides for the waiver of labour market tests for three categories of businesspersons and the waiver of the requirement of an employment authorization for a fourth category of business visitor seeking entry into Canada. NAFTA only applies to *temporary* entry. It does not affect the requirements for permanent residency in Canada. Temporary entry is generally considered to be a period of up to five years.

The four categories of businesspersons recognized under NAFTA are: A) Business Visitors; B) Traders and Investors; C) Intra-Company Transferees; and D) Professionals. Entrants in each of these categories are required to qualify for entry under Canadian health, safety, and security requirements and to indicate which category is applicable. Advice should be obtained with respect to specific legal requirements and procedures to satisfy compliance by entrants in these various categories.

Canadians seeking clearance for entry into the United States or Mexico should seek similar advice in those countries.

CHAPTER THIRTY-SIX

LAWS RELATING TO EMPLOYMENT

FEDERAL LAWS

Consistent with the constitutional division of powers between the federal and provincial governments, federal labour legislation governs labour relations and employment matters where industries and undertakings of an interprovincial, national, or international nature are concerned. For example, transportation, communications, and any work or undertaking that is for the general advantage of Canada, or for the advantage of two or more of its provinces, are federal matters.

The most significant of the federal statutes is the Canada Labour Code.[1] The Code covers three general areas of employment law. One part of the Code sets out minimum employment standards; it deals with hours of work, overtime pay, minimum wages, holidays, vacations, maternity leave, bereavement leave, notice of termination of employment, and unjust dismissal of non-unionized employees. Another portion of the Code deals with safety of employees. (The Code permits an employee to refuse to work where there is an imminent danger to the employee's health and safety; the Code also describes proper procedures to be followed when such refusal occurs.) The third part of the Code covers relations between trade unions and employers. The Code establishes the Canada Labour Relations Board, which administers provisions of the Code relating to the certification of unions as bargaining agents for employees. The Board also investigates claims of unfair labour practice and illegal strikes, and establishes procedures for lawful strikes and lockouts.

[1] R.S.C. 1985, c. L-2

Canada Pension Plan

Under the Canada Pension Plan, an employer is required to deduct from an employee a percentage of the employee's earnings from pensionable employment and remit that amount to the federal government together with an amount contributed by the employer.

Unemployment Insurance

An employer is required to deduct and remit unemployment insurance premiums on behalf of employees who work in insurable employment pursuant to the Unemployment Insurance Act (Canada). The employer is also required to pay a premium on behalf of its employees.

The amount of premiums to be deducted from an insured employee's salary, wages, or other remuneration is determined by use of the Unemployment Insurance Premium Tables issued by Revenue Canada, Taxation for the pay periods indicated. It is the obligation of every employer paying remuneration to a person employed in insurable employment to deduct the amount of premium payable and remit it together with the employer's premium to the Receiver General of Canada.

Employment Equity

The federal government passed the Employment Equity Act in 1986 in an effort to promote equity and non-discriminatory practices in the workplace. This legislation includes steps to reasonably accommodate members of certain target groups. It requires all federally regulated companies to do essentially two things: implement employment equity and report annually to the Minister of Human Resources and Development.

In addition, the federal contractors program applies to all companies in Canada (not just federally regulated employers) with 100 employees or more who bid on federal government contracts for goods or services in excess of $200,000. All such companies must certify that they will implement an employment equity plan that meets specified criteria.

PROVINCIAL LAWS

Each province has enacted provincial labour laws. Provincial laws deal with relations between trade unions and employers; labour standards (such as maximum hours of work, overtime, minimum wages, holidays, and notice of termination of employment); worker's compensation, occupational health and safety, and so on. Provincial labour statutes have established boards to administer labour-relations legislation. In Alberta, for example, the Board of Industrial Relations administers labour standards and labour relations. In Ontario, the Ontario Labour Relations Act[2] is administered by the Labour Relations Board; labour standards are administered by the Employment Standards Branch of the Ministry of Labour.

PROVINCIAL STATUTES — ONTARIO

The Employment Standards Act[3]

The Employment Standards Act establishes minimum terms and conditions of employment. The Act was designed to protect non-union employees; but all employees are covered, with certain limited exceptions. The Act prescribes requirements in many areas: maximum hours of work, payment for overtime, minimum wages, public holidays, vacations with pay, equal pay for male and female employees performing equal work, employee benefit plans, pregnancy and parental leave for employees, individual and group notice of termination of employment, and severance pay.

The Act also provides that where an employer sells a business or part of a business and the vendor's employees are retained by the purchaser, the employees are not deemed terminated by the sale. The period of employment with the purchaser includes the period of employment with the vendor for purposes of notice of termination, vacations, holidays, and pregnancy leave. For example, suppose an employer purchases a business or any part of a business; suppose the employer retains the employees for only one day, and then decides to terminate them. The notice of termination and severance pay must be based upon the period of employment with the vendor.

The Act sets out minimum requirements only. Any term or condition of employment included in a written or oral contract of employment that provides a greater right or benefit prevails over the minimum requirement imposed by the Act.

---

[2] R.S.O. 1990, c. L-2
[3] R.S.O. 1990, c. E-14

The Act does not affect an employee's civil remedies against an employer. For example, suppose an employee is hired for an indefinite period of time; suppose there is no written contract stipulating the notice period required to terminate the services of the employee. The common law requires that the employer give the employee reasonable notice of termination (or pay instead of reasonable notice) unless the employee is terminated for cause. Common law might provide an employee with a greater period of notice than does the Act. At common law, the period of notice depends upon an employee's position, length of service, age, and the availability of similar employment having regard to the employee's experience, training, and qualifications. An employer might terminate an employee in accordance with the Act, yet the employee could still sue the employer at common law. The employee might claim to be entitled to more notice, or might ask for pay in lieu of notice.

For an employee with more than ten years' service, the maximum notice of termination required under section 40 of the Act is eight weeks. Employers with a payroll of $2.5 million or more must also pay severance pay based on one week for each year of service, if the employee has five years service, to a maximum of 26 weeks. For an employee who does not exercise supervisory duties and who does not possess any high degree of technical skill, the notice and severance pay requirements set out in the Act may be sufficient to satisfy the common-law requirement of "reasonable notice." However, for management and professional employees, the period of notice required pursuant to common-law decisions will likely exceed the statutory minimum. Generally, managers and professional employees are entitled to from three to twenty-four months' notice.

An example is a 1961 Ontario case, *Lazarowicz* v. *Orenda Engines Limited*.[4] A forty-nine-year-old professional engineer with three years' service was awarded the equivalent of three months' salary. A more recent example is the 1979 case, *Blakely* v. *Victaulic Company of Canada*.[5] A fifty-four-year-old professional engineer, who rose to become the senior officer of a Canadian subsidiary of a U.S. company, was asked to move to the United States after twenty-four years of service with the company in Ontario. The employee treated the direction to move as a constructive dismissal. He sued and recovered twenty-one months' salary in lieu of reasonable notice of termination.

[4]   26 D.L.R. (2d) 433
[5]   [1979] 1 A.C.W.S. 351

The Workers' Compensation Act[6]

The Workers' Compensation Act provides for an employer-financed accident fund for medical aid and for loss of earnings. As well, the fund compensates permanent or partial disability caused by personal injury, accident, or illness that results from employment. The concept of fault does not apply: compensation is provided even for injuries sustained by an employee through the employee's own negligence. The rights under the Act replace the common-law right to sue the employer.

For purposes of payment into the fund, employers are divided into classes; the classifications depend upon the hazards in each industry. The Workers' Compensation Board establishes rates of payment for each class annually; rates are based on the cost of accidents occurring in the industries in each class. Some industries are not covered. For example, in industries where all employees work in offices, accidents on the job are infrequent; and coverage is often not provided.

When an accident occurs, an employer is required to fill out an accident report. The forms are provided by the Board. The report must be sent to the Board within three days of the accident.

The Workers' Compensation Board, which administers the Act, has very broad jurisdiction. Its decisions are protected from review by any court except on very limited grounds. However, there are several levels of appeal within the Board itself.

The Health Insurance Act[7]

Under the Act, the Ontario government has established a scheme of health insurance for all residents in Ontario. The scheme is called The Ontario Health Insurance Plan (OHIP) and is funded by an Employer Health Tax, paid by the employer as the name suggests.

Smoking in the Workplace

In Ontario, the Smoking in the Workplace Act prohibits smoking in any "enclosed workplace" with the exception of areas used primarily by the public or designated smoking areas. Employees who exercise their right to complain under this Act are protected

---

[6]   R.S.O. 1990, c. W. 11
[7]   R.S.O. 1990, c. H. 6

from intimidation, coercion, discipline, suspension, or dismissal. The Act does not override smoking restrictions in municipal by-laws unless the by-law is less stringent than the Act. Similar legislation to control smoking at the workplace and on common carriers under federal jurisdiction has also been enacted. In Quebec, the applicable legislation regulates the use of tobacco in certain public places with a view to ensure better protection for the health and well-being of non-smokers.

The Human Rights Code[8]

The Human Rights Code is of general application to all persons in Ontario.

It establishes that every person has a right to equal treatment with respect to employment, without discrimination because of race, ancestry, place of origin, colour, ethnic origin, citizenship, creed, sex, sexual orientation, age, marital status, family status, record of offences, or handicap. The Act prohibits any person from infringing the rights of others as established by the Code either directly or indirectly and either intentionally or unintentionally. The Code further establishes that every employee has the right to be free from sexual harassment in the workplace. The Code is administered by the Ontario Human Rights Commission, which investigates complaints and appoints Boards of Inquiry where appropriate.

The Occupational Health and Safety Act[9]

Employee safety is covered by the Occupational Health and Safety Act.

The Act provides that a health and safety committee is required at most workplaces where twenty or more workers are employed. The committee must consist of at least two persons; at least half the members of the committee must not exercise managerial functions; and at least half the members must be selected by the workers they represent, or by a trade union. The committee must meet at least once every three months. Where nineteen or fewer workers are employed, the Minister of Labour is authorized to require an employer, constructor, or group of employers to establish a health and safety committee. Where a committee is not required, a worker health and safety representative must be appointed by and from the workers.

[8]   R.S.O. 1990, c. H. 19
[9]   R.S.O. 1990, c. O. 1

The Occupational Health and Safety Act also contains provisions that give an employee broad powers to refuse to perform work where the employee has reasonable cause to believe that the machine or device being used, or the workplace in which the employee is working, is unsafe. The Act is administered by the Ministry of Labour, which conducts workplace investigations.

The Act sets out the duties of an owner, constructor, employer, supervisor, and worker. Breach of those duties can lead to charges and prosecution. The maximum fine for corporations is $500,000 for each offence.

The Ontario Labour Relations Act

Pursuant to The Ontario Labour Relations Act, every person is free to join a trade union of his or her own choice and to participate in its lawful activities. The Act provides for the certification of unions; it places constraints upon what an employer may and may not do when faced with a union-organizing campaign. For example, the Act prohibits an employer or persons acting on the employer's behalf from interfering with the selection of a trade union by the employer's employees. The employer may express views as long as the employer does not use coercion, intimidation, threats, promises, or undue influence. Once the trade union has signed up sufficient employees, it files an application for certification with the Ontario Labour Relations Board. The Board will then decide what is an appropriate bargaining unit. For example, the Board will exclude persons exercising managerial functions; the Board will determine the number of cards signed by employees in the bargaining unit. The Act provides that professional engineers are entitled to a bargaining unit composed exclusively of professional engineers, unless a majority of the engineers wish to be included in a unit with other employees.

Usually, if the union has signed up less than 40 percent of the employees in the bargaining unit, the application will be dismissed. If the trade union has signed up more than 55 percent of the employees, the trade union will be certified as the bargaining agent for the employees. If the trade union signs up between 40 and 55 percent of the employees, the Board will order a vote by secret ballot to determine whether a majority of the employees wish to be represented by the trade union.

Sometimes an employer is found to be in breach of the Act during the organizing campaign. The Board may certify the trade union as bargaining agent for the employees without a vote, even though the trade union has not signed up a majority of the employees.

The Act also provides procedural requirements governing collective bargaining between the employer and the union and strikes and lockouts. The procedures include conciliation and mediation, if necessary.

Pursuant to the Act, no employees shall strike and no employer shall lock out his or her employees while a collective agreement is in operation or until the employer and the union have completed the conciliation process.

The Act provides for compulsory arbitration by an independent arbitrator in certain cases: where an employee is disciplined or discharged; where the other party is not complying with the terms of the collective agreement; or where there is disagreement as to the interpretation of the collective agreement.

The Act prohibits an employer from engaging in "strike-related" misconduct. This is defined as engaging in unfair conduct intended to interfere with any right, under the Act, during the course of or in anticipation of an unlawful strike or lockout. This section also expressly prohibits an employer from retaining the services of professional strike-breakers.

On construction projects, it is important for the engineer to be aware that many general contractors and subcontractors are covered by collective agreements, some of which apply on a province-wide basis covering many different employers. Disputes can arise where union and non-union contractors are working on the same job site or where one union is on strike and picketing and other unions are not. Disputes can also arise where different unions are claiming the same work as work coming within their jurisdiction. Procedures have been established to resolve these disputes at the Labour Relations Board.

APPENDIX

COMMENTARY AND SAMPLE CASE STUDIES FOR ENGINEERING LAW COURSES AND EXAMINATION PROGRAMS

COMMENTARY

This book is used as a reference text in connection with the professional practice examination in engineering law and professional liability of Professional Engineers Ontario ("PEO"). It is also used in connection with engineering law and related courses at a variety of Canadian universities and colleges. This third edition provides an opportunity for some commentary intended to assist engineers in their preparation for, and writing of, engineering law exams. (The author has been involved in setting and marking the PEO's engineering law and professional liability examinations since 1981.)

The approach taken in this chapter, following this introductory commentary, is to provide further sample case studies that represent previous PEO examination questions and questions developed in the course of the author's teaching programs at the University of Toronto. Included at the end of the relevant chapters are representative answers to at least one of the case studies in each of the primary areas of relevance — contract law, business law, and tort (or negligence). As the engineer will be quick to appreciate, when it comes to the application of legal principles in the areas of tort law liability or the enforceability of clauses limiting liability, for example, once the fundamental principles are understood and demonstrated, their application to other factual situations should be straightforward. The inclusion of additional cases is intended to assist the engineer in becoming familiar with the types of questions to expect on an engineering law exam. Essentially, the process is one of problem solution, an area for which the exceptional analytical ability and training of engineers makes them particularly well suited.

For the purpose of studying engineering law and preparing for the PEO engineering law examination, it is advantageous to the engineer to take an engineering law course. The advantages of an instructor's insights and classroom participation are obvious. However, engineers in more remote locations may be unable to do so. With that group of engineers particularly in mind, the following general comments on some of the history, and recommended approach to examination preparation, are offered.

For many, if not most, engineers writing law exams as part of their application for licensure (or at the undergraduate level), the engineering law exam is the first law exam that they will have written. Accordingly, the prospect may be initially worrisome. However, provided the appropriate study is undertaken for the engineering law exam, the engineer shouldn't have cause to worry about the outcome. The PEO engineering law exam (and presumably other similar exams) is intended to test the awareness of engineers of the fundamental legal principles relevant to engineering. It is obviously not intended to create lawyers out of engineers.

Some historical perspective on the results of the PEO engineering law exam may be helpful. As mentioned, this particular examination program was implemented in 1981. Many examination candidates have demonstrated an excellent grasp of the subject and most of the examination candidates have successfully passed the engineering law examinations without difficulty, as should be the case. However, over the years, the failure rate has continued to hover around 10 percent. This, in the author's view, is much higher than it should be. The problem of the unsuccessful examination candidate lies primarily in inadequate preparation.

A continuing analysis of the reasons for failure reveals a pattern. Approximately 25 percent to 30 percent of the examination candidates who fail the exam do so because they are unable to effectively communicate. This inability of the examination candidate to effectively and clearly express himself or herself typically indicates a language difficulty that can be overcome with appropriate further language study. In making this point, it is emphasized that the engineering law exam is not a language examination. Identifying papers where candidates have difficulties in expressing themselves clearly enough is relatively straightforward for the marker.

More surprising is the fact that 70 to 75 percent of the candidates who fail the engineering law exam do so simply because they have failed to adequately prepare. The problem in attempting to rely solely on "common sense" and "general knowledge and experience" is that the nature of the subject matter is sufficiently technical that lack of appreciation of the legal principles quickly becomes apparent to the marker. As indicated at the outset of this text, an effort has been made to minimize the use of technical legal terms. However, relevant areas such as the essential criteria for tort liability, the technical basis upon which contractual obligations may sometimes be avoided, and issues relating to the enforceability of contract provisions that limit liability are examples of areas where an understanding of the application of technical fundamentals is required. So also is the mandatory "definitions" question, which is, in most cases, a ready source of half the marks a successful candidate needs. For example, the appropriate definition of "secret commission" (as a bribe or kickback, as described in the text) is a quick source of five marks on the examination. But this is not so if the candidate defines secret commission, as has been the case, as "a secret committee set up by the government for secret purposes, the recommendations of which will never be made public"; or as a "committee for the registration of trade secrets." Similarly, the definition of "common law" as used in this text refers to the theory of precedent or judge-made law, and an answer that identifies "the relationship between a man and a woman who live together who are not married" will fall short of the definition expected.

In answering the case type questions that typically set out a fact situation for analysis and application of legal principles, the following general comments are made to assist the candidate in formulating an answer:

1. The Facts   In the interest of time efficiency, the facts should not be restated by the examination candidate. A careful reading of the facts should suffice. A brief diagram, to identify contractual relationships, is a good idea, particularly in tort cases for the purpose of analyzing which parties may be potentially liable even though no contractual relationship may exist between the plaintiff and the defendant.

In addition, it is important to avoid assuming additional facts, if at all possible. In this regard, the case facts are often modified summaries or simplifications from actual cases. The intention is to provide a straightforward set of facts from which the important legal issues can be identified and to which the fundamental legal principles can be applied. It is not intended to provide any technical information that may be difficult to understand from an engineering perspective. Neither is it intended to provide a factual situation that generates a discussion of technical engineering principles. All that is intended is to provide a straightforward fact situation in order that the candidate can demonstrate awareness and application of the relevant fundamental legal principles. Accordingly, assumption of additional facts should be avoided as that usually unnecessarily complicates the case and the ability of the candidate to answer it in the time available.

2. Give Reasons   In answering, it is important to communicate to the marker an understanding of the relevant legal principles. Accordingly, these should be briefly stated in the course of identifying the issues and applying the fundamental principles to determine a logical outcome or conclusion. For example, in a tort case, the purpose of tort law should be stated and the essential criteria should be identified. As well, the approach the courts will apply to determine if the requisite standard of care was performed in the circumstances should be identified. Then the fundamental principles should be applied to the facts, leading to a reasoned conclusion. Sometimes candidates are quick to set down conclusions without reasons and the marker is unable to assess if the candidate understands the legal principles. Accordingly, an appropriate brief explanation is important.

Unlike engineering exams, the outcome of the analysis on the facts given may well lead to the possibility of more than one answer. This is the "judgment" that is ultimately necessary on the matter in dispute. For the purposes of the engineering law examination, a reasonable discussion of the issues, the arguments, and a reasonable conclusion, on the basis of the limited facts given, is all that can be expected.

3 Format   Length of answer and format is entirely up to the candidate. Given time constraints, point form answers are quite acceptable. Engineers and engineering students have a great deal of experience in allocating available time to the number of questions that need to be answered. In this regard, the PEO engineering law exam includes a mandatory definitions question (defining five out of eight or nine terms for 25 marks); plus three additional questions of equal value to be answered (out of four or five additional questions given). The case questions typically include opportunities for the candidate to answer a tort question, a contracts question and, often, a question that relates to legal issues arising for an engineer in business or as a contract administrator.

4. Reference to Decided Cases   As already indicated, the purpose of this text is to familiarize engineers with legal concepts and principles relevant to the engineering profession with a view to assisting the engineer to hopefully avoid legal problems and to identify problems on a timely basis should they arise. For PEO examination purposes, it is not essential that the engineer memorizes any case names or statute references. An exception is in the area of "fundamental breach" where briefly referencing the *Harbutt's Plasticene* and *Syncrude* v. *Hunter* cases is an effective and efficient means of identifying an understanding of the case developments that have occurred in this area. Generally, however, reference to specific cases is an impressive "embellishment" of the answer, but is not essential. What is important is to analyze the facts, identify and state the applicable legal principles, and come to a reasoned conclusion. Similarly, specific statute references are not at all necessary for the purposes of the PEO engineering law exam. If the engineer needs to refer to the provisions, for example, of the Construction Lien Act in the course of the engineer's practice, he or she can do so as appropriate. However, there is no need to memorize statute references for the purposes of the PEO engineering law exam.

5. Important Chapters for PEO Engineering Law and Professional Liability Examination For PEO examination purposes, candidates should focus their studies in particular on fundamental concepts and principles described in Chapters 1 and 2, Chapters 4 through 24, and Chapters 28, 30, 32, 33 and 36.

Hopefully, these comments will assist the examination candidate and alleviate unnecessary concerns. Provided the candidate has studied the materials and can appropriately communicate an appreciation for the fundamentals, the examination candidate should not find the law exam any unreasonable hurdle to licensure as a professional engineer! For most candidates, succeeding on the engineering law exam is, and should be, a very straightforward process.

ADDITIONAL SAMPLE CASE STUDIES

Bearing in mind the foregoing commentary on engineering law exams, the following cases are intended to assist readers and examination candidates in becoming familiar with the types of examination questions that can be expected. The approach that should be taken in setting out a satisfactory answer has been illustrated earlier at the end of the relevant chapters in this text.

SAMPLE TORT CASES

1 Provincial Life of Ontario Inc. ("Provincial"), an insurance company, retained an architect to design a new corporate head office in North York, Ontario. Provincial, as client, and the architect entered into a written client/architect agreement in connection with the project. According to the agreement, the architect was to prepare the complete architectural and engineering design for the project.

In order to carry out the structural engineering aspects of the design, the architect engaged the services of a structural engineering firm. The architect and the structural engineering firm entered into a separate agreement to which Provincial was not a party.

To determine the nature of the soil on which the project would be constructed, two shallow test pits, each about 1.25 metres deep, were dug on the site at locations selected by the architect. The architect telephoned the structural engineering firm's vice-president and requested that the firm send out a professional engineer to examine the soils exposed in the test pits.

Based on information received from the professional engineer sent to examine the soil, the vice-president of the structural engineering firm reported to the architect that the test pits had revealed a silty clay. The vice-president also recommended to the architect that a soils engineer be engaged to carry out more thorough and proper soils tests. The architect rejected the recommendation stating that there was not "enough room in the budget" for more soils tests.

The architect succeeded in persuading the vice-president to send a letter to Provincial giving a "soils report" based on the examination of the shallow test pits. The vice-president stated in a letter to Provincial, that based on its examination of the test pits, the soil was a fairly uniform mixture of clay and silt, which would be able to support loads up to a maximum of 100 kPa.

The structural engineering firm then completed its structural engineering design on the basis of the maximum soil load reported to Provincial.

The project was constructed in accordance with the plans and specifications. Subsequently, the building suffered extensive structural change, including severely cracked and uneven floors and walls.

On the basis of an independent engineering investigation by an engineer retained by Provincial, it was determined that the extensive structural change in the building had resulted from the substantial and uneven settlement of the building. The investigation also determined that the subsoil in the area of the building consisted of 30 to 40 metres of compressible marine clay covered by a surface layer of drier and firmer clay two metres in depth. The investigation also revealed that the test pits that were dug had not penetrated the surface layer into the lower layer of compressible material.

What potential liabilities in *tort law* arise from the preceding set of facts? Please state the essential principles applicable to a tort action and apply these principles to the facts above. Indicate a likely outcome of the matter.

2 Ontario Industrial Laundry Inc. ("OILI") is the owner of several laundry plants in Ontario. OILI's operations include handling the laundry for various industrial and institutional facilities around the province. OILI decided to build a large new plant in Brampton. The new plant would replace a number of smaller and aging facilities OILI operated nearby.

OILI engaged an architectural firm, Clever and Really Useful Design Developments Inc. ("CRUDDI"), and entered into an architectural services agreement with it. Under the agreement, CRUDDI was to design the new plant and prepare construction documentation necessary to build it. According to the agreement, CRUDDI was to design "the most modern and technically up-to-date laundry in Canada."

CRUDDI hired a number of engineering consultants to provide the various engineering design services necessary for the project. Of these, Mechanical Engineering Systems and Services Inc. ("MESSI") was to design the air conditioning and handling system.

Although MESSI did not have a contract with OILI, it worked closely with a representative of OILI who specified that, as it was important to provide comfortable working temperatures in the plant, the air conditioning and handling system must be able to provide working temperatures in the range of 22° to 25° C and a minimum of 18 air changes per hour.

OILI, on the basis of competitive tenders, awarded the contract for the construction of the new plant to Dominion Industries and Related Technologies Inc. ("DIRTI"). The contract price was $15 million. DIRTI completed the construction in accordance with the contract drawings and specifications.

Almost immediately after having commenced its operations in the new plant, OILI experienced problems in the air conditioning and handling system. The temperature in the working areas was excessive, reaching 38° C in the summer months. In the compressor room, the temperature reached 50° C and caused malfunctions. In addition, the circulation was poor and the air quality was offensive. The employees began suffering fatigue and other ailments and it became necessary for them to take frequent "heat breaks."

CRUDDI and MESSI tried several times to remedy the problems but they were unsuccessful. OILI retained Top Industrial Designs Inc. ("TIDI"), another mechanical engineering company, to conduct an independent investigation. TIDI determined that the air conditioning and handling system was underdesigned. The air conditioner's chilling unit had a capacity of only 230 tonnes; a larger unit having a capacity in the order of 600 tonnes should have been specified. In addition, the exhaust and intake vents on the roof were located too close to each other and caused exhausted air to re-enter the plant.

TIDI determined that the system would require $1.1 million in modifications in order to meet the plant's specifications. It also indicated that, had the system been specified and constructed as it ought to have been in the first place, construction costs incurred by OILI would have been $400,000 higher, that is, $15,400,000.

What potential liabilities in *tort law* arise in this case? In your answer, explain what principles of tort law are relevant and how each applies to the case. Indicate a likely outcome to the matter.

3 A contractor specializing in farm buildings was engaged by an owner to design and construct a barn to be placed over a manure pit. The contract between the contractor and the owner provided that the contractor would be responsible for both the design and construction of the barn and manure pit.

The contractor had previously designed and built barns over manure pits, but had never designed a manure pit of the size and shape required by this owner. In preparing the design, the contractor contacted an engineer who was employed by the Department of Agriculture of Ontario (the "government engineer"). The government engineer was a government employee and not a consulting engineer. However, the government engineer was employed by the government to assist farmers and contractors to work out their plans and, although the government engineer never received any payment from the contractor, the government engineer had previously provided advice to the contractor in connection with the design of farm buildings.

The contractor and the government engineer never met to discuss the plans but discussed the matter by telephone. Eventually, the contractor left a copy of the plans on the government engineer's desk and the government engineer reviewed the plans and forwarded the following handwritten message to the contractor:

> "Good set of plans. I like the detail.Wish I could spend that amount of time on each project. Keep up the good work."

After the manure pit was constructed in accordance with the plans, the walls of the manure pit cracked badly and had to be rebuilt.

The owner sought the advice of another engineer (the "consultant") who redesigned the manure pit prior to the remedial construction taking place. The consultant noted that the contractor's plans had two particular deficiencies:

1. The plans showed the reinforcing rod to be in the middle of the wall. The rebar should have been closer to the inside of the wall for maximum support.

2. There was a complete absence of any rebar schedule on the plans.

The consultant concluded that the original plans were deficient insofar as the structural steel components and requirements vital to the integrity of the concrete wall were missing.

When the owner discovered that the contractor had sought advice from the government engineer, the owner sued both the contractor and the government engineer on account of the extra costs incurred by the owner in having the manure tank redesigned and reconstructed.

The government engineer's position was that the government engineer really hadn't carefully reviewed the plans, but had simply looked through them and hadn't really understood that the contractor was seeking advice on the detailed sufficiency of the plans.

Do you think the owner would be successful in a *tort* claim against the government engineer? In your explanation, discuss the tort law principles that a court would apply to determine whether and to what extent the government engineer would be liable.

4 Ace Furniture and Appliance Distributors Inc. ("ACE") is a company primarily engaged in the distribution of household appliances and furniture in Ontario and Quebec. In order to improve its distribution facilities, ACE decided to build a new warehouse in Markham, Ontario to replace two smaller warehouses that it was currently leasing.

ACE engaged Super Projects Engineering Consulting Services Inc. ("SPECS"), an engineering company, and entered into an engineering services agreement with it. Under the agreement, SPECS was to prepare the design of the new warehouse and to provide inspection services to ensure that the warehouse was constructed in accordance with the plans and specifications for the project.

ACE, on the basis of competitive tenders, awarded the contract for the construction of the warehouse to King Construction Ltd. ("KING"). During the course of construction, KING contracted with Queen Equipment Supply Ltd. ("QUEEN"), an equipment supplier, to provide an operator and equipment to backfill the foundation as well as a truck to haul away any extra fill material. In order to perform the backfilling operation, QUEEN supplied a backhoe and one of its employees, Phil Scooper, to operate the backhoe.

During the backfilling, one of the construction workers on the site noticed cracks appearing in the warehouse's foundation and quickly motioned to Phil Scooper to stop the backfilling because there was a problem.

On closer examination, it turned out that there were several severe cracks in two of the foundation walls. As a result of the cracking, ACE incurred extra costs in repairing the foundation. In addition, the project was delayed for four weeks.

ACE retained an independent engineering consultant to investigate the cause of the cracking. Based on its review of the project documents, the consultant determined that SPECS had instructed KING to commence backfilling after only two days of curing time had elapsed following the pouring of the concrete foundation. The consultant's report indicated that the National Building Code of Canada recommended a minimum period of curing of seven days and that no such requirement was included in SPECS's plans and specifications.

The report also indicated that according to several witnesses on the job site, Phil Scooper was dumping fill material directly into the trench beside the foundation while the backhoe's bucket was high in the air. The report indicated that the operator should have pushed the fill materials into the trench and that the operator's procedure of "high dumping" of the fill material did not conform with the general practice of the operator's trade. In the consultant's opinion, high dumping was a factor in the cracking.

What potential liabilities in *tort law* arise in this case? Explain how the essential principals of tort law apply to this case. Indicate a likely outcome to the matter.

5 Mammoth Undertaking Ltd.("Mammoth"), a development company, retained the firm of Sharpe Architects ("Sharpe") to design a six-storey office building. Sharpe also agreed with Mammoth that Sharpe would provide or arrange for inspection services during the course of construction of the project in order to ensure that construction was carried out in accordance with the project plans and specifications.

Sharpe prepared a conceptual design and retained Abel Engineering ("Abel") to prepare the detailed structural design for the project and also to carry out inspection services to ensure that all structural aspects of the construction of the project were carried out in accordance with the project plans and specifications.

Abel prepared the structural design and eventually Mammoth awarded the contract for the construction of the project to a general contractor, Swift Construction Ltd.("Swift").

Abel appointed one of its employee engineers, James Newman, a recent engineering graduate, as Abel's representative and inspector on the construction site.

Construction commenced during the month of October and soon thereafter Swift recommended to Mammoth that a substantial cost savings could be effected if the specified fill material around the foundation was changed to a more readily available material. Mammoth sought Sharpe's advice on the suitability of the proposed alternative fill material and indicated to Sharpe that it was most important that a decision be made as soon as possible in order to complete as much of the foundation and backfilling as possible prior to frost conditions setting in.

Sharpe, in turn, referred the matter to Abel through its representative Newman, requesting that Abel approve the proposed change as quickly as possible in the circumstances. Newman determined that the original fill material had been specified by an engineer who no longer worked for Abel and that the specification had been made on the basis of a careful investigation of soil conditions at the site. Newman contacted one of Abel's vice-presidents and was authorized to advise Sharpe as to the suitability of the alternative fill material after conducting an appropriate investigation.

Under pressure from both Mammoth and Swift to approve the proposed fill material without delaying the construction schedule, Newman approved the change of materials without giving due consideration to the possible repercussions.

The substitute material did not drain as well as the material originally specified; in fact, it retained some water and, as it expanded during freeze up, it caused significant cracking in the foundation walls, necessitating remedial work resulting in substantial additional expense being incurred by Mammoth. In addition, the completion of the project was considerably delayed as a result.

Explain the potential liabilities *in tort law* arising from the preceding set of facts. In your explanation, discuss and apply the principles of tort law and indicate a likely outcome of the matter.

6 National Stores Inc. ("NATIONAL"), the owner of a grocery store chain in Ontario, contracted with an architect to design and prepare the construction documentation for a new store in Kenora, Ontario.

The architect produced some general construction specifications that included a requirement that an automatic sprinkler system, conforming to the National Fire Protection Association ("NFPA") standards, be installed.

The architect retained an engineering firm pursuant to a separate agreement to which NATIONAL was not a party. Under the contract the engineering firm was to prepare the detailed engineering design for the project, including the sprinkler system. The engineering design was to conform to the architect's general specifications.

The design of the sprinkler system was prepared by a recent engineering graduate employed by the engineering firm. Not being familiar with the NFPA requirements, the employee read certain sections of the standards but did not have enough time, given other project responsibilities, to pay close attention to all the details. The employee's completed sprinkler system design was reviewed by a professional engineer. Although the professional engineer did not perform a detailed check, the professional engineer considered the design satisfactory.

Six months after the store opened for business, a fire occurred early one morning. The fire caused substantial damage to the store and to its inventory and NATIONAL had to close the store for repair.

NATIONAL retained a consulting engineer to conduct an independent investigation. The consulting engineer determined that the sprinkler system was inadequately designed. Specifically, the design did not conform to the NFPA standards, which required, among other things, that the coverage per sprinkler head was not to exceed 10 square metres. The engineer determined that 10 percent of the sprinkler heads were designed to cover an area as high as 25 square metres. The report indicated that, in the engineer's expert opinion, had the sprinkler head spacing conformed to the NFPA standards, the fire should have been quickly extinguished and would not have spread to any great extent.

What liabilities in *tort law* may arise in this case? In your answer, explain what essential principles of tort law are relevant and how each principle applies to the case. Indicate a likely outcome of the matter.

7 A developer/owner retained an architect to design an office tower complex in downtown Toronto. In the agreement between the developer/owner and the architect the architect agreed to be responsible for all aspects of design of the complex, including all structural, mechanical, and electrical engineering design aspects.

The architect entered into a contract with a mechanical engineering firm for all mechanical engineering design services for the project, particularly the heating, ventilating, and air conditioning systems.

The complex was designed and ultimately constructed at a cost of $125 million.

The air conditioning system as designed and specified by the mechanical engineering firm did not perform satisfactorily, as evidenced by start-up and performance tests. Major design modifications and alterations to equipment already installed had to be undertaken at additional project costs in excess of $2 million before the air conditioning system performed satisfactorily and the project could be completed. As a result, the completion date of the project occurred two months later than scheduled.

The developer/owner, when initially faced with the air conditioning performance shortfalls, retained a second mechanical engineering firm to investigate the reasons for the problem. The second mechanical engineering firm prepared an opinion report for the developer/owner which concluded that the employee engineer of the mechanical engineering firm that prepared the design had made significant errors in the design calculations that resulted in the deficient performance of the air conditioning system. The opinion report also stated that the suppliers of the air conditioning equipment had complied with the specifications included in the project contract documents.

What potential liabilities in *tort law* arise in this case? In your answer, explain what essential principles of tort law are relevant and how each principle applies to the case. Indicate a likely outcome to the matter.

8 J. Smart, P. Eng., was retained by a municipality in Southern Ontario to design and supervise the construction of a bridge.

Smart and the municipality executed a contract for Smart's design and supervisory services.

Smart estimated that construction of the bridge would cost the municipality approximately $1.75 million. The municipality pointed out to Smart that budgetary restrictions were such that it would not be economically feasible for it to proceed with construction if the cost were to exceed $1.8 million.

Smart entered into a contract with a firm of soils experts, Acme Underground Ltd., to advise Smart on sub-surface conditions at the site. Acme Underground Ltd. was made fully aware that its services were being requested in connection with the bridge design. On the basis of its subsurface investigations, Acme Underground Ltd. reported to Smart that no difficulties whatsoever should be encountered with subsurface conditions, insofar as all drill holes indicated that the footings could easily be designed to rest on bedrock.

Smart completed the detailed design, the plans and specifications were finalized, and the construction of the bridge ultimately awarded to ABC Construction Limited at a cost of $1.65 million. The municipality entered into a contract with ABC Construction Limited as general contractor for the project. The form of the contract had been prepared and approved by Smart.

In excavating, ABC Construction determined that the subsurface conditions were not as represented in the plans and specifications. Indeed, only two-thirds of the footings could be placed on bedrock at the design elevations. Another firm of soils experts, Subsurface Wizards Inc., was called in to investigate and ultimately concluded and reported that the Acme Underground employee who was responsible for the initial investigations had not drilled enough test holes to accurately predict the nature of the subsurface conditions.

After extensive additional test borings were carried out, a revised design of the structure was prepared, which included the driving of piles and a new footing design to ensure a secure basis for the foundation. These changes in design resulted in an extra cost of $350,000 being requested by ABC Construction, much to the annoyance of the municipality.

Smart determined that ABC's price of $350,000 for the extra work was a reasonable price in light of the revised design.

What potential *tort* liabilities arise from the preceding set of facts? In identifying the potential liabilities in tort law, explain the application of tort law principles to the facts as given. Indicate a likely outcome of the matter.

9 A municipality, as owner, retained an architect to design a new police station. The architect entered into a contract with an engineering firm to perform structural design services in connection with the project.

In performing soils investigations, the engineering firm's employee engineer assigned to the project examined two shallow test pits and recommended to the architect that proper deep soils tests be taken. However, the architect rejected the engineer's recommendation, informing the engineer that expensive soils tests were not part of the owner's budget for the project.

The engineer submitted a "soils report" to the owner on the basis of the superficial examination of the shallow test pits. Neither the architect nor the engineer indicated to the owner that the engineer had recommended to the architect that a more thorough subsurface investigation be undertaken.

The design of the police station was completed and the building was constructed in accordance with the project drawings and specifications.

Within twelve months of completion of the engineering design services the new police station "settled" very badly on one side and extensive remedial foundation work was necessary to correct the settlement problems.

Upon investigating the reason for the settlement problems, another consulting engineering firm concluded that the design should never have proceeded without the more detailed and thorough subsurface investigation that the original project engineer had recommended to the architect.

What potential liabilities arise from the preceding set of facts? In identifying the potential liabilities in *tort law*, explain the application of tort law principles to the facts as given. Indicate a likely outcome of the matter.

10

Creative Developments Ltd. ("Creative") entered into a contract with Towers & Associates, Architects, for the design and preparation of contract documentation necessary to construct a twenty-storey office building of unique design. Towers & Associates also agreed to perform inspection services during the construction of the office building.

Towers & Associates prepared a conceptual design, and Biggar Inc., an engineering consulting firm, agreed to prepare the detailed structural design for the project.

The Biggar firm was retained by Towers and Associates. However, it appeared that the firm's very busy schedule would not permit its senior design engineer sufficient time to attend to the design personally. Biggar turned the matter over to one of its employee engineers, Hilary Abel, a recent engineering graduate in whom the Biggar firm had great confidence.

Abel completed the design and Biggar's senior design engineer reviewed it. Although not having checked all of Abel's calculations in detail, Biggar's senior design engineer concluded that the design appeared satisfactory and the senior design engineer's professional engineer's stamp was affixed to the design drawings. The drawings were submitted to Towers & Associates. Towers & Associates included Biggar's structural-design drawings in the contract documents for the construction of the project.

Creative entered into a construction contract with Sound Construction Ltd. to erect the office building. During the course of construction, the partly finished building collapsed, resulting in considerable damage to Creative's property. As well, there was a substantial delay in the completion of the office building. Creative conducted an investigation as to the cause of the collapse, and they obtained the opinion of another consulting engineering firm. The second firm thought that the structural design, as supplied by the Biggar firm, was inadequate in the circumstances.

The investigation also revealed that Towers & Associates had retained Subdata Inc., soils experts, to carry out soils tests, prior to Abel's preparation of the structural design. Creative subsequently obtained an opinion from another firm specializing in soils tests; the second firm concluded that many more tests should have been performed by Subdata Inc. As well, the second firm thought that the results of the tests performed were "borderline": the employee of Subdata Inc., who prepared the original report, had very seriously erred in concluding that the test results were satisfactory. The second soils experts also concluded that the subsurface conditions resulted in serious settlement on one side of the building during construction; that settlement contributed to the collapse.

Explain the potential liabilities arising from the preceding set of facts *in tort law*. Discuss, with reasons, a likely outcome of the matter.

SAMPLE CONTRACT LAW CASES

11 An owner and a contractor entered into a written contract for the construction of a $20 million chemical plant in Sarnia, Ontario. The contract provided that the plant would be constructed in accordance with the plans and specifications that had been prepared by the owner's engineering consultant. Under the contract, the owner, through the engineering consultant, was permitted to make changes to the design of the plant with the amount payable to the contractor being adjusted accordingly. However, the contract further provided that the contractor could not proceed with any change in the work without a written order signed by the owner and that no claim for additional compensation on account of a change would be valid without such a written order.

As the work progressed, the engineering consultant certified payments for amounts due to the contractor on the basis of the amount of work performed during each month. Several of the monthly payments included additional compensation for extra work performed by the contractor on account of relatively minor changes to the design of the plant. In total there were 55 such minor changes. In each case, the contractor had proceeded with the extra work and was paid additional compensation despite the fact that no written order was given by the owner authorizing the extra work or the additional compensation.

During the course of the work, the engineering consultant made a major change to the design of the plant. It was anticipated that the change would require an additional $1.7 to $2.0 million of work by the contractor and would require four months to complete. The contractor requested the owner's approval before proceeding with the extra work. The owner indicated orally to the contractor that the contractor should proceed with the work and that a written order authorizing the change would be issued once the details of the design change were finalized.

The contractor commenced performing the additional work for the major design change in January, 1990 and invoiced the owner on a monthly basis. Although the owner never did issue a written order authorizing the additional work, the contractor was paid for the extra work that was performed in January, February, and March of 1990. The contractor completed all of the extra work in April 1990 and submitted an invoice for payment, which included $950,000 for extra work performed in April.

The owner refused to pay the $950,000 on the basis that no written order by the owner was given authorizing any extra work, as required by the written contract. Is the contractor entitled to the $950,000? Explain.

12 An owner and a contractor entered into a written construction contract which provided that payments were to be made by the owner to the contractor within five days subsequent to an engineer's certificate being issued and that, if the owner should fail to pay the contractor within such five-day period any sums that the engineer has certified as due, the contractor would be entitled to terminate the construction contract. The contractor had been the lowest bidder on the project and, as the construction proceeded, became concerned that, because of its low bid, the contractor would lose money on the contract.

During the first two months of the six-month construction schedule, the engineer certified payments due by the owner to the contractor and such payments were made to the contractor within five days of such certification.

At the end of the third month of construction, the engineer certified a further sum as due to the contractor. In spite of having received the engineer's certificate, the owner requested that prior to payment the contractor obtain the corporate seal of one of its subcontractors on a document supporting the engineer's certificate. The contractor promised to obtain such corporate seal. However, the contractor never did obtain the corporate seal; the five-day payment period passed and ten days later the contractor notified the owner that it was terminating the contract on account of the owner's failure to pay within the five-day period pursuant to the terms of the construction contract.

Was the contractor entitled to terminate the contract in the circumstances? Explain.

13 An owner and a contractor entered into a written construction contract. According to the contract, the contractor would be paid a lump sum price to construct a factory. The contractor was to complete the work by August 30, 1992.

The contract provided that if the owner delayed the contractor in performing the work, the contractor would be entitled to additional time to complete the work and to be reimbursed by the owner for reasonable costs incurred by the contractor as a result of the delay. Section 4.3 of the contract provided:

> *Section 4.3*: No claim for a time extension or for costs shall be made for delay unless a written notice describing the delay is given to the owner not later than 14 calendar days after the commencement of the delay.

Under the contract, the owner was to purchase and supply specialized manufacturing equipment that the contractor was to install in the factory. The owner was to arrange for and have the equipment delivered to the site by March 15, 1992. The owner delivered the equipment on March 22, 1992 and the contractor immediately commenced to install it. At a meeting at the site between the owner and the contractor on March 23, 1992 to discuss the status of the work, the contractor verbally indicated that it had incurred additional labour and equipment rental costs as a result of the owner's delay in delivering the manufacturing equipment. The contractor explained that because the manufacturing equipment was not available as scheduled, the contractor had incurred 8-person days in additional labour costs and two days' in rental costs for a crane. The contractor said that it would be seeking compensation for these additional costs. The owner assured the contractor that it would be paid for the delay and asked the contractor to provide the owner with a detailed written statement of the additional costs "within the next month or so."

On April 10, 1992, the contractor provided the owner with a detailed written statement indicating additional costs of $9,400 as a result of the delay. On May 10, 1992, the owner responded to the contractor's detailed statement and indicated that the contractor was not entitled to the additional costs claimed because the contractor had failed to give the owner a written notice of delay within the time required by section 4.3 of the contract.

Is the contractor entitled to the additional costs claimed? Explain.

14 A contractor, ABC Construction Co. ("ABC"), submitted a bid on a construction project. ABC's bid price of $6 million was very low in comparison to the other bidders. In fact, the three other bidders had each bid amounts in excess of $7 million.

The contract was awarded to the lowest bidder. The contract conditions expressly entitled the contractor to terminate the contract if the owner did not pay monthly progress payments within ten days following certification by the project engineer that a progress payment was due.

ABC commenced work on the project and soon determined that it would likely suffer a major loss on the project, as it had made significant judgment errors in arriving at its bid price. ABC also learned that, in comparison with the other bidders, ABC had "left a million dollars on the table."

After the fourth monthly progress payment was certified as due by the project engineer, ABC was approached by the owner for additional information relating to bills from an equipment supplier, the cost of which comprised a portion of the fourth progress payment amount. The owner requested that the additional information be provided prior to payment of the fourth progress payment being due. Although the signed contract did not obligate ABC to obtain such additional information, a representative of ABC verbally informed the owner that ABC would provide the additional information. However, ABC never did so.

Eleven days after the progress payment had been certified for payment, ABC notified the owner that ABC was terminating the contract as the owner had defaulted in its payment obligations under the specific wording of the contract.

Was ABC entitled to terminate the contract? Explain.

15 Arbour Pulp & Paper Company ("ARBOUR") entered into a written equipment supply contract with Recovery Exchangers and Turbines Inc. ("RECOVERY"). According to the agreement, RECOVERY was to design, manufacture, and deliver a heat recovery steam generator to ARBOUR's pulp and paper mill in Ontario for a purchase price of $3.5 million. ARBOUR would arrange to install the equipment in its mill as part of a co-generation system for the purpose of converting steam into electricity.

According to the agreement, RECOVERY was to begin manufacturing the equipment on February 1, 1992 and deliver the finished product to ARBOUR on or before March 30, 1993. The agreement provided that ARBOUR would pay the $3.5 million purchase price in monthly instalments over the manufacturing period. The agreement contained the following provision:

> Each instalment of the purchase price shall become due and payable by ARBOUR on the last day of the month for which the instalment is to be made. If ARBOUR fails to pay any instalment within 10 days after such instalment becomes due, RECOVERY shall be entitled to stop performing its work under this contract or terminate this contract.

As the work progressed, RECOVERY invoiced ARBOUR for each monthly instalment. Although ARBOUR paid the first instalment on time, it was more than 20 days late in paying each of the second, third, fourth, fifth, and sixth instalments. RECOVERY never once complained about the late payments, even when ARBOUR apologized for the delayed payments and commented in meetings with RECOVERY that ARBOUR's current cash flow difficulties, resulting from the impact of recessionary times, were the reasons for the late payments.

By the middle of September 1992, it became apparent to RECOVERY that due to serious cost overruns resulting from its own design errors and lack of productivity, it would stand to lose a substantial amount of money on the contract by the time the equipment would be completed. Although the instalment for August had been invoiced and was due on August 31, 1992, ARBOUR had not yet paid it by September 15, 1992. On September 15, 1992, RECOVERY terminated the contract.

Was RECOVERY entitled to terminate the contract? Explain.

16 Clearwater Limited, a process-design and manufacturing company, entered into an equipment-supply contract with Pulverized Pulp Limited. Clearwater agreed to design, supply, and install a cleaning system at Pulverized Pulp's Ontario mill for a contract price of $200,000. The specifications for the cleaning system stated that the equipment was to remove 95 percent of prescribed chemicals from the mill's liquid effluent in order to comply with the requirements of the environmental control authorities in the area in which the mill was located.

In addition, the signed contract between Clearwater and Pulverized Pulp contained a warranty provision whereby Clearwater stated it would, for a period of one year from the date of installation, repair defects in the process and equipment arising from faulty design or parts or workmanship. But Clearwater accepted no responsibility whatsoever for any indirect or consequential damage arising as a result of the contract.

The cleaning system installed by Clearwater did not meet the specifications. In fact, only 70 percent of the prescribed chemicals were removed from the effluent. As a result, Pulverized Pulp Limited was fined $10,000, and was shut down by the environmental control authorities. Clearwater, having then been paid $185,000 of the $200,000 contract price, made several attempts to remedy the situation by altering the process and cleaning equipment, but without success.

Pulverized Pulp eventually contacted another equipment supplier. For an additional cost of $250,000, the second supplier successfully redesigned and installed remedial process equipment, that cleaned the effluent to the satisfaction of the environmental authorities, in accordance with the original contract specifications between Clearwater and Pulverized Pulp.

Explain and discuss what claim Pulverized Pulp Limited can make against Clearwater Limited in the circumstances.

17 A contractor agreed to design, supply, and install large holding tanks for storing and dispensing heavy wax liquefied under heat for a manufacturing process. A clause in the contract between the factory owner and the contractor limited the contractor's "total liability for loss, damage, or injury to the factory owner resulting from the performance of work by the contractor to an amount not to exceed the value of the contract price, being $450,000," and also provided that under no circumstances would the contractor be responsible for any indirect or consequential damages, however caused.

The contractor designed a system wholly inadequate for the purpose intended and was careless in testing it. In fact, as a direct result of the contractor's negligent design and testing procedures a fire resulted that totally destroyed the entire factory together with two adjacent buildings belonging to third parties with total damage assessed at $4.5 million.

Discuss and explain the extent of the contractor's liability to the factory owner and the neighbouring property owners in this situation.

18 ACE Construction Inc. is a company primarily engaged in the business of supplying heavy equipment used in construction. As part of the company's economic plan to expand its business, ACE became interested in the rock crushing industry.

ACE had become aware that International Metals Company Ltd. ("IMCO") required a contractor to crush, weigh, and stockpile approximately 250,000 tonnes of ore. As ACE believed this was an excellent opportunity to venture into the rock crushing business, it decided to tender on the IMCO contract.

In order to tender on the contract, ACE set out to purchase the necessary equipment to crush the material. ACE was contacted by a representative of Rock Busters Ltd., a company that sold such equipment. After visiting the IMCO site and determining the nature of the material to be crushed, the representative discussed the IMCO contract with ACE. After performing a number of calculations, the representative determined and guaranteed that the equipment Rock Busters would provide would be capable of crushing the material at a rate or 175 tonnes per hour. On the basis of the guarantee, Rock Busters and ACE entered into a contract. Rock Busters agreed that if ACE were successful in its tender to IMCO, Rock Busters would provide the equipment for a price of $400,000. The contract also contained a provision limiting Rock Busters' total liability to $400,000 for any loss, damage, or injury resulting from Rock Busters' performance of its services under the contract.

Based on the information provided by the representative, ACE prepared and submitted its tender to IMCO. IMCO accepted the tender and entered into a contract with ACE to crush the material.

The rock crushing equipment was set up at the IMCO site by employees of Rock Busters and crushing operations commenced. However, from the beginning there was trouble with the operation. One of the components of the crusher, called the cone crusher, consistently became plugged by the accumulation of material.

Each time the cone crusher became plugged, the operation would have to be shut down and the blockage cleared manually. In some cases, such blockages caused damage to the equipment. Rock Busters made several unsuccessful attempts to correct the defect by making modifications at the site and at its factory. The crushing equipment was never able to crush more than 30 tonnes of materials per hour.

In order to meet its obligations under the IMCO contract, ACE hired another supplier to correct the defects in the Rock Busters equipment. For an additional $500,000 the supplier replaced the cone crusher with one manufactured by another company. The modified equipment was able to crush the material at the rate of 180 tonnes per hour. The total amount paid by ACE to Rock Busters was $350,000.

Explain and discuss what claim ACE can make against Rock Busters in the circumstances. Would ACE be successful in its claim? Why? In answering, please include a summary of the development of relevant case precedents.

19 Supercleen Limited, a manufacturing company, entered into an equipment supply contract with Red Fire Mines Limited. Supercleen agreed to design, supply, and install a dust collection system at Red Fire Mine's northern Ontario smelter for a contract price of $400,000. The specifications for the dust collection system stated that the dust collection equipment was to remove 95 percent of prescribed exhaust particles from the exhaust gases in order to comply with the requirements of the environmental control authorities.

In addition, the signed contract between Supercleen and Red Fire Mines also contained a provision limiting to $400,000, Supercleen's total liability for any loss, damage, or injury resulting from Supercleen's performance of design, supply, and installation services to Red Fire Mines pursuant to the contract.

The dust collection system as installed by Supercleen did not meet the specifications. In fact, only 60 percent of the prescribed exhaust particles were removed from the exhaust gases. As a result, Red Fire Mines was faced with the threat of substantial fines and possible shutdown by the environmental control authorities. Supercleen refused to remedy the defective equipment without being assured of compensation from Red Fire Mines of any costs in excess of $400,000 incurred in connection with such remedial work.

At the time of discovering that the system failed to meet the specifications, Supercleen had already received $350,000 from Red Fire Mines, and Red Fire Mines refused to pay anything further to Supercleen.

Red Fire Mines contacted another equipment supplier who, for an additional cost of $500,000 successfully designed and installed remedial equipment sufficient to clean the exhaust gases to the satisfaction of the environmental authorities and in accordance with the original contract specifications between Supercleen and Red Fire Mines.

Explain and discuss what claim Red Fire Mines Limited can make against Supercleen Limited in the circumstances. In answering, please include a summary of the development of relevant case precedents.

20 Hyper Eutectoid Steel Inc. ("HESI") is a company that produces various types of steel for industrial applications. In order to increase the strength of its steel products, HESI uses a process of quenching and tempering. During the quenching stage, hot steel is quickly cooled with water. During the tempering stage, the steel is then heat treated for an appropriate time. The process requires large amounts of water and heat.

Faced with rising costs for energy, HESI decided to install a heat recovery system. The system would include a heat exchanger to recover heat from the cooling water in the quenching stage. The recovered heat, then, would be used to heat the steel in the tempering stage.

HESI entered into an equipment supply contract with Energy Recovery and Recyclings Systems Inc. ("ERRS"). ERRS agreed to design, supply, and install a heat recovery unit for a contract price of $600,000. After an analysis of HESI's processes, ERRS determined and guaranteed in the contract that the heat recovery system would recover 40 percent of the heat in the cooling water and that this would result in substantial savings in energy costs.

The contract also contained a provision limiting ERRS's total liability to $600,000 for any loss, damage, or injury resulting from ERRS's performance of its services under the contract.

The heat recovery system was installed and was operational; however, certain defects in the heat exchanger prevented the system from ever recovering more than 5 percent of the heat in the cooling water. After repeated unsuccessful attempts by ERRS to remedy the defects, HESI hired another supplier, who, for an additional $800,000, replaced the heat exchanger and was able to achieve the level of performance originally promised by ERRS. The total amount received by ERRS under its contract was $500,000.

Explain and discuss what claim HESI can make against ERRS in the circumstances. In answering, please include a summary of the development of relevant case precedents.

TABLE OF ABBREVIATIONS TO CASE AND STATUTE CITATION

CANADA

Reports
C.C.L.T. Canadian Cases on the Law of Torts
C.L.R. Construction Law Reports
C.P.R. Canadian Patent Reports
D.L.R. Dominion Law Reports
Ex. C.R. Exchequer Court Reports
O.L.R. Ontario Law Reports
O.R. Ontario Reports
O.W.N. Ontario Weekly Notes
S.C.R. Supreme Court Reports
W.W.R. Western Weekly Reports
A.C.W.S. All Canada Weekly Summaries

(Revised) Statutes
R.S.A. Alberta
R.S.B.C. British Columbia
R.S.C. Canada
R.S.M. Manitoba
R.S.Nfld. Newfoundland
R.S.N.S. Nova Scotia
R.S.O. Ontario
R.S.P.E.I. Prince Edward Island
R.S.S. Saskatchewan
R.S.N.B. New Brunswick
R.O.N.W.T. Northwest Territories
R.O.Y.T. Yukon

UNITED KINGDOM:

App. Cas. Appeal Cases
All E.R. All England Reports

E.R. English Reports
A.C. Appeal Cases
W.L.R. Weekly Law Reports
C.A. Court of Appeal
K.B. King's Bench Division
P.D. Probate Division
Exch. Exchequer
R.P.C. Reports of Patent Case
Ch. Chancery Division

TABLE OF CASES

INDEX